£20

ANTI-RACIST SCIENCE TEACHING

ANTI-RACIST SCIENCE TEACHING

EDITED BY DAWN GILL AND LES LEVIDOW

'an association in which the free development of each
is the condition of the free development of all'

FREE ASSOCIATION BOOKS · LONDON · 1987

First published in Great Britain by
Free Association Books
26 Freegrove Road
London N7 9RQ

This collection and introduction
© Dawn Gill and Les Levidow 1987

The individual authors retain the copyright
in their articles

British Library Cataloguing in Publication Data

Anti-racist science teaching.
 1. Science – Social aspects – Study and
 teaching – Great Britain
 I. Gill, Dawn II. Levidow, Les
 306'.45'07041   Q175.5
 ISBN 0-946960-63-1
 ISBN 0-946960-64-X Pbk

Typeset by Rowland Phototypesetting Ltd,
Bury St Edmunds, Suffolk
Printed and bound in Great Britain by
Short Run Press Ltd, Exeter

CONTENTS

Notes on Contributors   vii
Acknowledgements   x
Addresses   xi

General Introduction   1
Dawn Gill and Les Levidow

RACISM IN SCIENCE
Introduction   14
Racist Society, Racist Science   16
Robert M. Young
Racism in Scientific Innovation   43
Les Levidow
Sickle Cell Anaemia: An 'Interesting Pathology'   59
Michael G. Michaelson
Interpreting the Production of Science   76
Robert M. Young

ANTI-RACIST CURRICULUM CHANGE
Introduction   90
Nutrition and Hunger: Two Classroom Approaches   94
Liz Lindsay
Biology Teaching in a Racist Society   107
Michael Vance
Multicultural versus Anti-Racist Science: Biology   124
Dawn Gill, Europe Singh and Michael Vance
Kenya: The Conservationists' Blunder   136
Malcolm Green
Science Curriculum Innovation at Holland Park School   147
Dawn Gill, Vinod Patel, Ashok Sethi and Henry Smith

vi   Anti-Racist Science Teaching

RACE
Introduction   176
Pseudo-Scientific Racism   178
Peter Fryer
Are the Races Different?   198
Richard Lewontin

ASSESSMENT
Introduction   210
GASP! A Critique   213
Grazyna Baran, Les Levidow and Birendra Singh
Graded Assessments: Hijacking 'Process'   219
Birendra Singh
'Ability' Labelling as Racism   233
Les Levidow

BHOPAL
Introduction   270
Mass Death at Bhopal: Whose Responsibility?   271
Barbara Dinham
Union Carbide's Double Standards   280
Barry Castleman
Bhopal: Backward or Advanced?   283
Tara Jones

OFFICIAL DOCUMENTS AND COMMENTARIES
Introduction   294
ILEA Anti-Racist Statement   296
ILEA Correspondence   301
The Hargreaves Report: Monocultural Education?   307
Science 5–16: The DES Holds the Line   312

Index   319

NOTES ON CONTRIBUTORS

Grazyna Baran studied physics at the University of Kent, taught science at several London secondary schools, and now teaches adult education at Morley College.

Barry Castleman works as a public health consultant and chemical engineer in Baltimore, Maryland. He is author of *Asbestos: Medical and Legal Aspects* (1984).

Barbara Dinham works for the Transnationals Information Centre, London. She is co-author of *Agribusiness in Africa* (Earth Resources Research, 1983).

Peter Fryer became interested in the problem of racism when he covered, as a young reporter, the arrival of 500 Jamaican settlers on the *Empire Windrush* in 1948. He is author of *Staying Power: The History of Black People in Britain* (Pluto, 1984).

Dawn Gill has taught geography in ILEA schools since 1971. She is co-editor of the journal *Contemporary Issues in Geography and Education*. Seconded to the ILEA anti-racist strategies team between 1983 and 1986, she has worked on curriculum development in a range of subjects.

Malcolm Green trained as a teacher and biologist. He has worked in Kenya on the problem of rodent pest control, and in both Britain and Cameroon as a biology teacher. He is currently working on countryside interpretation in the Newcastle area.

Tara Jones has been active in opposing toxic industry and toxic waste dumps in Ireland. He is author of *Managing Mass Murder: From Bhopal to Chernobasel* (forthcoming).

Les Levidow is an unAmerican who moved to London in 1976. Formerly a school science teacher, he now teaches the politics of science in adult education and at Middlesex Polytechnic. He has edited several collections on social aspects of science, technology and medicine; he was Managing Editor of the Radical Science Series and now of *Science as Culture* at Free Association Books.

Richard Lewontin is a population geneticist at the Comparative Museum of Natural History at Harvard University. A long-standing member of Science for the People, he is co-author of *The Dialectical Biologist* (Harvard, 1985).

Liz Lindsay, now teaching at Willesden College of Technology, was formerly Head of Biology at Quintin-Kynaston School, London, where she helped initiate the school's Multi-Cultural Anti-Racist Working Party in 1979. She is a member of the All London Teachers Against Racism and Fascism editorial collective and the Inner London Teachers Association Education Committee; she has written for many other anti-racist publications.

Michael G. Michaelson has been a member of the Medical Committee for Human Rights, a medical student and sociology graduate. He is author of *Healing: Notes on Medicine and Revolution*.

Vinod Patel lived in Kenya and a remote Indian village before arriving in Britain. While at the University of Sussex, he served as Education Officer in the Student Union. Later he was involved with a North London-based Asian community group and setting up Saturday morning mother-tongue classes. He has taught at Holland Park School since 1983.

Ashok Sethi, born in Kenya, came to Britain via India. He has been teaching at Holland Park School since 1977. In 1986 he went on secondment, based at the Institute of Education, to develop anti-racist/multicultural science teaching methods.

Birendra Singh is at present Head of Science at a comprehensive school in Redbridge, Essex. Previously he taught science and physics in comprehensive schools in Newham and Islington (London) for twelve years.

Notes on Contributors   ix

Europe Singh has taught maths in ILEA schools for fifteen years. Between 1983 and 1986 he was seconded to ILEA's anti-racist strategies team. He is co-ordinator of the ACD's Campaign for Anti-Racist Mathematics.

Henry Smith, at present a head of faculty at Holland Park School, has taught there since 1969. Formerly he taught science at further education institutions. He is author of the prize-winning book *Amazing Air* (Methuen, 1985), which has been translated into ten different languages.

Michael Vance, a black biology teacher, has taught in Tottenham and Hackney (London) since 1978. He is a member of the Inner London Black Teachers Group and the Haringey Black Pressure Group on Education. He is involved with the Broadwater Farm Defence Campaign and the Haringey Independent Police Committee.

Robert M. Young grew up in racially segregated Texas. He studied philosophy at Yale, medicine at Rochester, and history and philosophy of science at Cambridge, where he taught for many years. He left to write and to do political and editorial work in London. He founded Free Association Books and is Editor of *Free Associations* and *Science as Culture*. He is author of *Darwin's Metaphor* and other studies in the politics of nature, human nature and science.

ACKNOWLEDGEMENTS

Reprinted Articles

Michael Michaelson, 'Sickle Cell Anaemia: An "Interesting Pathology"', copyright 1971 by Noah's Ark, Inc. (for *Ramparts* magazine), is reprinted from the October 1971 issue, pp. 52–58, by permission of Bruce W. Stilson.

Robert M. Young, 'Interpreting the Production of Science', first appeared in the 29 March 1979 issue of the *New Scientist*, London, the weekly review of science and technology, and is reprinted with their permission.

Peter Fryer, 'Pseudo-Scientific Racism', is adapted from his book, *Staying Power: The History of Black People in Britain*, London, Pluto Press, 1984, by permission of the author and publisher.

Richard Lewontin, 'Are the Races Different?', is reprinted from *Science for the People*, vol. 14, no. 2 (1982), pp. 10–14, by permission of the author and magazine.

Barbara Dinham, 'Mass Death at Bhopal', is reprinted from *International Labour Reports*, no. 8 (1985) with her permission.

ILEA, 'Anti-Racist Statement', is excerpted with the permission of the Inner London Educational Authority.

Illustrations

Brent Sickle Cell Centre, London, p. 74; **Holland Park School**, p. 149, p. 155; **Beverly Chorbajian**, p. 177; **Tana Acton**, p. 200; **Secondary Examination Council**, London, p. 223, p. 230; **Bhopal Disaster Monitoring Group**, Japan, p. 278.

ADDRESSES

All-London Teachers Against Racism and Fascism (ALTARF), 16 Panther House, Mount Pleasant, London WC1. Tel. (01) 278 7856.

Association for Curriculum Development (ACD), PO Box 563, London N16 8SD. Tel. (01) 254 9323.

Association for Curriculum Development in Science (ACDS), c/o Science Department, Holland Park School, Airlie Gardens, London W8. Tel. (01) 727 5631.

British Society for Social Responsibility in Science (BSSRS), 25 Horsell Road, London N5. Tel. (01) 607 9615.

Contemporary Issues in Geography and Education (CIGE), Allan Wells International, Farndon Road, Market Harborough, Leics. LE16 9NR. Tel. (0858) 34567.

Earthscan, 10 Percy Street, London W1. Tel. (01) 580 7574.

Free Association Books, 26 Freegrove Road, London N7 9RQ. Tel. (01) 609 5646/0507.

Greenpeace, 36 Graham Street, London N1 8LL. Tel. (01) 608 1461.

International Labour Reports, 300 Oxford Road, Manchester M13 9NS.

Links, c/o Third World First, 274 Banbury Road, Oxford OX2 7DZ. Tel. (0865) 723823.

New Internationalist, 42 Hythe Bridge Street, Oxford OX1 2EP. Tel. (0865) 728181.

Oxfam, 274 Banbury Road, Oxford OX2 7DZ. Tel. (0865) 56777.

Race & Class, c/o Institute of Race Relations, 2–6 Leeke Street, King's Cross Road, London WC1. Tel. (01) 837 0041.

Radical Science Journal/Series, c/o Free Association Books.

Sage Race Relations Abstracts, 28 Banner Street, London EC1. Tel. (01) 253 1516.

Science as Culture, c/o Free Association Books.

Science for People (SfP), c/o BSSRS.

Science for the People (SftP), 897 Main Street, Cambridge, MA 02139, USA. Tel. (617) 547 5580.

Socialist Environment & Resources Association (SERA), 9 Poland Street, London W1. Tel. (01) 439 3749.

Third World Network, c/o Consumers Association of Penang, 87 Jalan Cantonment, Penang, Malaysia. Tel. 373511.

Transnationals Information Exchange (TIE), 9 Poland Street, London W1. Tel. (01) 734 5902.

War on Want, 37 Great Guildford Street, London SE1. Tel. (01) 620 1111.

GENERAL INTRODUCTION

DAWN GILL AND LES LEVIDOW

Recent initiatives to develop anti-racist education, both in Britain and in other countries, have barely touched science teaching. Unfortunately, most teachers find it difficult to imagine how such proposals would apply to science, seen as 'objective' knowledge. The few teachers who have tried to develop anti-racist approaches have met with accusations that they are illegitimately 'dragging politics into the schools'.

This book is about the politics of science education. It analyses how racism permeates science and science teaching. In so doing, it locates both racism and science in their wider political and economic context. The aim is to involve science teachers in the process of exposing racist ideology and challenging racist practice.

Although the examples here come mostly from Britain, they can be taken as a paradigm case relevant elsewhere as well. Because of Britain's former role as a leading colonial power, aspects of its school system have been adopted – voluntarily or otherwise – by many other countries.

RACISM

Racism is far more than conscious, or even unconscious, kinds of personal prejudice, though these clearly perpetuate racist institutional practices. It is more, too, than the way institutions operate to the disadvantage of black people. The state racism implicit in British nationality laws, for example, clearly rein-

2 Anti-Racist Science Teaching

forces the personal prejudices of white people while making black people more vulnerable to economic exploitation.

Racism, however, has its roots in the development of colonialism and imperialism; it is integral to the global economy. A fuller understanding of racism comes only from analysing that history and its legacies in the social and economic relationships between black and white people today. In the search for profits, Western powers have historically enslaved black people, conquered and impoverished much of the Third World, and induced millions of its people to emigrate to metropolitan countries. There they have been treated as cheap labour – an 'internal colony' to be exploited or expelled as expedient.

As regards the role of modern Western science, its research priorities and applications have been orientated towards im-

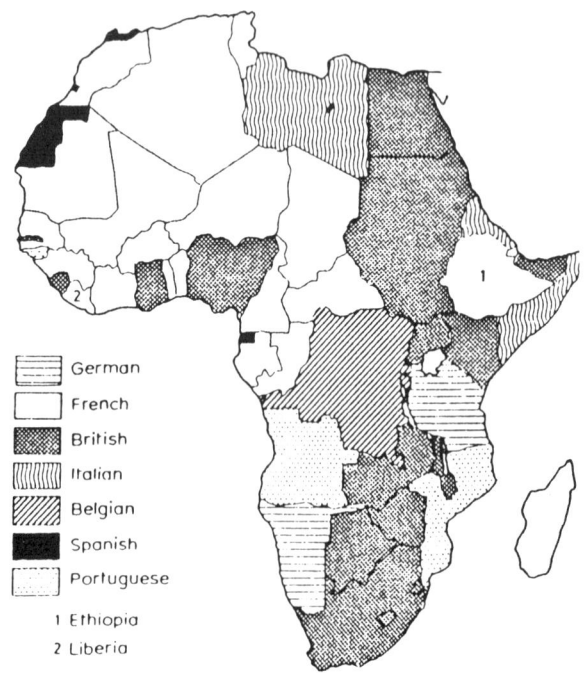

Roots of racism in Europe's division of Africa

posing military and economic power. Racial stereotypes, especially those given legitimation by science, can be understood and undermined only in relation to the history of imperialism, still in the making today.

Western racism is part of the structure of Western capitalism and all its institutions, including science. Racist ideology and practice may operate at all levels of social, political and economic interaction, from the interpersonal to the international. Thus it is possible for people to feel that they are 'not racist', while perpetuating institutional racism through their involvement in an exploitative economic system or merely by teaching science in the mainstream curriculum.

SCIENCE TEACHING

Few science teachers would espouse blatantly racist views. Most would claim that they treat all children alike, regardless of colour or creed. However, science serves a racist society in many subtle ways. An uncritical teaching and learning of science as currently practised inevitably engages the teacher and learner in maintaining structural racism.

Indeed, science teaching embodies a subtle form of racist propaganda which is all the more difficult to detect because science is commonly perceived to be politically neutral.

In particular, science teaching
☐ masks the real political and economic priorities of science;
☐ hides its appropriation of non-Western scientific traditions;
☐ often attributes people's subordination or suffering to nature – be it biological or geographical factors – rather than to the way science and nature itself have been subordinated to political priorities;
☐ is permeated by an ideology of race, both racist in origin and racist in effect;
☐ plays a key part in an exploitative economic and political system;
☐ perpetuates assumptions about nature and human nature that support inequality; and
☐ is an alienating experience for many students.

4 Anti-Racist Science Teaching

Furthermore, predominant methods of science teaching and student assessment select for those students most 'able' to compete in rituals of note-taking, memorization and individual timed tests – all of which help to define divisions among students according to 'ability level'. Racism, sexism and social class are key determinants in the 'ability' labelling process, a fact clearly revealed in examination results. This labelling process, together with the content and pedagogies of science education, add up to a system of indoctrination: that is, science education presents science as politically neutral and presents one viewpoint to the exclusion of others, while discouraging any critical questioning of the interests which science serves and the uses to which it is put.

Education should involve the examination of alternative viewpoints, and alternative explanatory frameworks. Thus the predominant model of schooling cannot be described as education in any true sense of the word.

The notion that science education is politically neutral is in itself highly political, and should be recognized as such. Further, the purposes served by this myth of neutrality need to be clearly illuminated. For too long, critically aware teachers have been denounced as subversives who are politically motivated, as if the standard school curriculum were not so. The politics of the status-quo curriculum must be challenged, so that public debate focuses on that curriculum rather than on those who criticize it.

STATE POLICY

The world-view reflected through school curricula – and the science curriculum in particular – is rooted in Britain's colonial past and its global economic interests today. It is this perspective which underpins racism, sexism and class inequality.

When, in 1985, the Department of Education & Science (DES) announced its new policy for teaching 'Science 5–16', it entirely ignored racism; in effect the policy provided a far stronger basis for opponents of anti-racist education than for its advocates. The then-Minister of Education, Sir Keith

Joseph, even asked parents to notify him if they thought that their children were being 'indoctrinated' at school: his speech ignored the racist or sexist indoctrination which is part of the status-quo curriculum.

The DES has been under right-wing pressure to go even further. An attack on anti-racist education has come from a Tory pressure group, the Monday Club, which has proposed that the DES issue guidelines to local education authorities and examination boards regarding multicultural education: 'This should not involve a fostering of ethnic minority cultures, nor a denigration of Britain's history, heritage or institutions.' (The choice of verb itself is telling: to denigrate or 'to blacken', presuming the negative quality of blackness.) Clearly the aim is to preserve the myth of Britain's imperial glory in 'civilizing' so-called primitive Third World peoples.

There are marked contrasts between central and local government policy. However, even where local education authorities have anti-racist policies, teachers who try to implement them encounter great difficulties. In 1983 the Inner London Education Authority (ILEA) became one of the first in Britain to adopt an anti-racist policy. At first the Authority was unable to offer guidance as to the structure and content of an anti-racist curriculum, although the policy gave teachers encouragement to develop such a curriculum. At the time there was great confusion between the terms 'anti-racist education' and 'multicultural education': these tended to be used interchangeably. 'Multiculturalism' was often suggested as an antidote to racism.

Some teachers – science teachers in particular – denied the need for anti-racist curriculum change. The ILEA's Staff Inspector for Science sent a letter to all heads of science departments, asserting that science was in effect 'multicultural' anyway, by virtue of possessing universal validity. His arguments, though in some ways idiosyncratic, nevertheless reflected the views held at the time by most science teachers, who considered both multicultural and anti-racist education as largely irrelevant to their practice.

Other teachers, already committed to anti-racist education,

6   Anti-Racist Science Teaching

continued to analyse the existing curriculum for racist bias and proposed ways of transforming the science curriculum.

Unfortunately, ILEA's official encouragement for anti-racist initiatives has been tempered by related practices. Not only has ILEA resisted student demands for protection from racist attacks, but teachers who supported one such student protest (at Daneford School) found that the head teacher responding by calling on the police . . . to arrest them.

We say: "GET POLITICS OUT OF THE SCHOOLS" Raise standards, not slogans!

Children in ILEA schools are being indoctrinated by left-wing propaganda. This has got to be stopped.

* Anti-police videos have been shown to kids

* The teachers' dispute is worse in London than anywhere else

* Striking miners gave lectures to children without anyone from the other side to balance the argument

* Spending on political handouts has rocketed - ILEA's Labour Leader now has a staff of three for her own personal publicity

* Many text books have been banned for loony far-left political reasons

This crazy approach means that better-off people are increasingly sending their children to private schools. This is Labour effectively dividing the rich from the poor - just like the Tories!

The Alliance promises to change all this once you've voted us in.

EDUCATION WILL COME FIRST ONCE MORE

General Introduction 7

And what of ILEA curriculum reform? Before becoming ILEA's Chief Inspector, David Hargreaves chaired a committee which recommended changes for *Improving Secondary Schools*. Despite its many apparently progressive proposals, the 'Hargreaves Report' – like the DES policy – hardly mentions institutional racism, and is based on a deficit model of underachievement. Further, it does not challenge racist ideology in the curriculum. The Authority has tended to capitulate to right-wing individuals and pressure groups raising such slogans as 'Get politics out of our schools! Raise standards, not slogans!' (SDP/Liberal Alliance ILEA election leaflet, 1986).

Indeed, the ILEA has recently set up a small 'watchdog body' – the 'Controversial Issues Committee' – of inspectors and others not noted for either their understanding of or commitment to anti-racist education, to vet curriculum materials which are considered 'politically sensitive'. This is the Authority's response to complaints from right-wing ideologues about peace studies and anti-racist education: these scrutineers argue that such curriculum content is 'political' and therefore should not be taught in school. Meanwhile, the racist content of materials used every day seems not to have attracted a passing glance.

The scrutiny of 'political' teachers and materials, and the media attacks on them, have been effective. Anti-racist educators have been accused of propaganda and their career prospects jeopardized. So much is this the case that many teachers now hesitate to implement the Authority's official policy. The practice of exposing racist ideology in the traditional school curriculum, and developing alternative materials, has somehow become defined as subversive activity, while perpetuating racist propaganda remains largely unchallenged.

HOW THIS BOOK WAS WRITTEN

As a result of the ILEA's anti-racist policy, a team of four advisory teachers was seconded to promote curriculum development. This team drew together a group of science teachers whose work had already begun to take account of racism within the science curriculum and who had begun working towards practical strategies for anti-racist education through science. The group made contact with members of the Radical Science Collective and All London Teachers Against Racism and Facism (ALTARF). Together they initiated a self-education group to develop the theoretical framework and practical classroom strategies for anti-racist education through science. This group included a variety of people involved in science education and writing, as well as classroom teachers.

Our first meeting, which took place in March 1984, addressed the question: 'How can science teaching encompass anti-racist education?' Background discussion documents had been written by group members on topics such as food and health, 'race' and racist ideolgy, and educational testing.

Discussion focused on how these related to three levels of racism and science:
1. blatant racism (as in concepts of 'racial' differences or hereditarianism);
2. patronage and priorities in science for setting the agenda of research and development; and
3. metaphysics and culture of science (why Western science treats objects as separate from their environment, and from social values).

We realized that we faced an extremely complex political and pedagogic issue, which encompassed the following:
☐ the permeation of racist ideology through the existing science curriculum;
☐ the operation of institutional racism through science as a practice and science as a school subject;
☐ institutional racism within the school as an organization and within the education system;
☐ the official image of science as neutral knowledge;
☐ science teachers' acceptance of that image, the teaching methods which promote it;
☐ an exam system geared in turn to that pedagogy; and
☐ pressure from central government and right-wing organiz-

ations against challenges to reactionary and racist practices and curriculum content.

Given the tentative character of our own alternative practices, we most needed to share our classroom materials and experiences and learn from each other how we could strengthen them. Only by evaluating our successes and failures could we eventually present our work in a form that might be useful to other teachers. We held regular discussions of written papers from group members, who often wrote several drafts as a result.

Some of the papers were presented by group members at a week-long series of discussions on 'Anti-Racist Education

Through Science', held in November 1985 in the Students' Union at the Institute of Education, University of London. The conference was originated by ILEA anti-racist advisory teachers, in the face of opposition from other officers of the Authority. Our science teachers' group – by that time the 'Association for Curriculum Development in Science' – sponsored the event and jointly organized it.

Three years after we began, we publish the result of our work. It is by no means the last word on the subject. We offer it as a contribution to the important debate on the political content of science education. The book's introductory section demonstrates that the science/politics distinction is a false one, by showing how science and technology embody values, and racist priorities in particular. It should not be surprising, then, that traditional school science courses perpetuate racist ideology and practices. Specific case studies present anti-racist approaches to biology, nutrition and wildlife conservation, as well as one school's experiment with reorganizing the curriculum across disciplinary boundaries. In particular the concept of 'race' itself is shown to be an artificial construct, originally devised to justify slavery and colonial policies. The Bhopal disaster, inflicted by the USA's Union Carbide Corporation on an Indian town, provides a stark example of institutional racism in operation. Finally, the book challenges the usual methods of student assessment, by which 'ability' labelling intensifies divisions along class, cultural and gender lines; it suggests ways to transform the pedagogy and assessment methods that end up labelling so many students as 'low-ability'.

This book, the first of its kind, proposes a theory and practice of anti-racist education through science. We anticipate that it will become central to the debate about racism in education. We hope that its publication will stimulate further analysis of the role that science plays in a racist society, further examination of the role that schooling plays in supporting racist economic and political structures, and further proposals for anti-racist science teaching.

THE ASSOCIATION FOR CURRICULUM DEVELOPMENT

The original science teachers' group, having begun to develop anti-racist strategies and resources, is keen to work with other teachers. We have recently set up the Association for Curriculum Development in Science (ACDS). The ACDS continues to hold regular meetings to discuss teachers' experiences and the wider politics of science teaching, such as our critique of the DES document, *Science 5–16*. The group provides its own in-service teacher education, as well as running seminars for educational institutions on request.

The ACDS is affiliated to similar groups in other subjects – in maths, geography, history, economics and primary education – to form The Association for Curriculum Development, an umbrella organization which promotes links between teachers involved in anti-racist and anti-sexist education throughout Britain and in other countries. It has working contact with radical teachers' groups in Australia, South Africa and Tigray.

The ACD promotes and co-ordinates opposition to racist and sexist propaganda through schooling, which has for too long been allowed to pass as politically neutral: thus we challenge the status-quo content of the curriculum which characterizes the school systems of capitalist societies.

Racism in Science

INTRODUCTION

Racism in science teaching is far more than an issue of curriculum content or 'racist teachers'. The problem lies partly in the political priorities that guide science itself, and thus the questions that are pursued – or not – in research and development programmes. In that sense, racism in science can be understood as a special case of how actual science embodies values while maintaining a privileged status as supposedly value-neutral.

In this section Robert M. Young argues – contrary to popular images of 'the scientific method' – that commercial and military interests have shaped most of our science, technology and medicine. There is no science other than the science that gets done – which in turn depends upon the questions that scientists and funding institutions consider most important or interesting to ask and support. Using several examples, he shows the racist priorities which influence what counts as a problem for science to solve; indeed, racism runs as deeply as the basic assumptions and concepts of science. He goes on to criticize teachers' resistance to acknowledging those political aspects of science; he concludes by suggesting how anti-racist science teaching would need to challenge official definitions of what science is.

Les Levidow connects racism in science to the wider role of science and technology in providing more effective ways of exploiting labour and natural resources. Using examples of population control, agriculture, energy and microelectronics, he shows how scientific innovations are geared to power relations in the global capitalist economy. Thus the racist effects of science can be understood to result from the way research and development priorities intersect with the economic domination of black and Third World people.

Illustrating the wider arguments is a remarkable case study by Michael Michaelson. In *Ramparts* magazine (1971) he documented the problem of sickle cell anaemia, which has afflicted significant numbers of black people in the USA and Britain. The disease, largely neglected for a long time, was eventually taken up by the medical establishment as an 'interesting pathology', useful for advancing scientific careers. The official research seemed hardly designed to help those who were suffering the effects of the disease and other kinds of deprivation.

Lastly, Robert M. Young surveys the different ways in which commentators have analysed the social aspects of science, often through recognized disciplines and courses. Although this article was not originally written in relation to racism, it is relevant in several ways. It helps to deflate the pretensions of scientists who attempt to speak authoritatively outside the areas which they have seriously studied. It also introduces analytical tools useful for those who want to challenge the racist content of science and its hidden values more generally.

RACIST SOCIETY, RACIST SCIENCE

ROBERT M. YOUNG

The problem of racism might seem an eccentric starting point for rethinking a science curriculum. It would appear that the problems raised from this vantage point, however legitimate, would not take long to clear up, at least in principle. The ways in which race and IQ are discussed in the general culture, and the ways in which they are taught in some contexts, are obvious examples of potential 'abuses' of science. Taking the argument further, the concepts of race and IQ are themselves problematic and would disappear in a different society – one which did not concern itself with human differences in certain physical attributes or in the ordinal ranking of individuals according to the ability to think in certain abstract ways.

Even within the existing culture, it can be argued that 'race' has no biological foundation; it is only, at best, a statistical concept of relatively pure gene pools. No biologist could draw lines in a large mixed sample so as to demarcate sharply a given race from another. Research on blood antigens would provide a good way of addressing this question.

This approach could be contrasted with overtly racist writings by reputable scientists like C. D. Darlington, a Fellow of the Royal Society and a Professor at Oxford. His writings celebrate 'racial differences' and are overtly racist; examples are *The Evolution of Man and Society* and *The Little Universe of Man*. Consider the following quotations from his earlier book (1969):

By inbreeding within classes Irish society was thus genetically fixed and stabilized at a pre-industrial stage and this has hindered its evolution in step with its neighbours. Only the disappearance of the barriers between Catholic and Protestant can break this evolutionary stalemate. (p. 455)

Thus the slaves were now racially changed. They were now more variable in features and in colour, in intelligence and in temperament... The genetic basis of the original relation of master and slave had disintegrated. (p. 592)

In short, racial discrimination has a genetic basis with a large instinctive and irrational component. (p. 606)

All the great races of man differ in smell; they dislike one another's smell and are kept apart by it. (p. 645)

The Nobel Prize winner Konrad Lorenz was also an overt racist during the German Nazi era, when he wrote:

> Nothing is so important for the health of a whole *Volk* as the elimination of 'invirent types': those which, in the most dangerous, virulent increase, like the cells of a malignant tumour, threaten to penetrate the body of a *Volk* ... Especially today the great difference depends upon whether or not we can learn to combat decay phenomena, in *Volk* and in humanity, which arise from the lack of natural selection. In just this contest for survival or extinction, we Germans are far ahead of other culture-*Volks*. (quoted in Kalikow, 1978)

Another Fellow of the Royal Society, Nobel Prize winner and Oxford Professor, Sir Hans Krebs, argued that biology 'proved' that trade unionism was against nature. Indeed, he said, '... a continued decrease in working hours is an unrealistic and utopian dream. The survival of nations, alas, is a matter of ruthless competition with other nations' (quoted in Young, 1973).

Therefore, we do not need to look at the most obviously biased materials and the work of disreputable scientists in order to see the intermixture of scientific concepts and value systems. Indeed, one can find work by reputable anthropologists, economists and other social scientists which claim that nature 'proves' that right-wing political theories are true.

The measure of IQ, like that of race, links politics with the typing and ranking of people for elitist reasons. This has become obvious in the debate around the work of Arthur Jensen and Sir Cyril Burt, both of whom have recently been exposed by careful research (Kamin, 1974; Levidow, 1977). This kind of thinking has been linked to wider issues by the recent books of Martin Barker (*The New Racism*) and Alan Chase (*The Legacy of Malthus*).

The topics of race and IQ help us to move on to a deeper level, one which illuminates why the task of creating an anti-racist science curriculum is far from eccentric but leads us to the heart of science. As soon as we move off the question of race and IQ as abuses of science or as bias, we are faced with a more searching question, of which they are striking examples: where do scientists' questions come from? What leads to the priorities, agendas, assumptions and fashions of science? Science is not something in the sky, not a set of eternal truths waiting for discovery. Science is a practice. There is no other science than the science that gets done. The science that exists is the record of the questions that it has occurred to scientists to ask, the proposals that get funded, the paths that get pursued, and the results which lead curiosity to rest and scientific journals and textbooks to publicize the work.

My view is that the problem of racism in science teaching is a special case of this deeper issue. The agendas in scientific and technological research reflect the prevailing values of a given culture. Research and development are the embodiment of values in theories, therapies and things. A racist society will have a racist science. A different society could have different science and, indeed, could break down the convenient and confusing barrier between science and the rest of society. The course of my argument is an attempt to move from obvious examples to less obvious ones so that we can see the larger issue.

THE QUESTIONS THAT GET ASKED

Nature 'answers' only the questions that get asked and pursued long enough to lead to results that enter the public domain. Whether or not they get asked, how far they get pursued, are matters for a given society, its educational system, its patronage system, and its funding bodies.

Let us take another example in the general area of 'race'. There is a disease which is specific to certain peoples, one group of whom come from a certain part of Africa. It causes episodes of anaemia due to a genetic defect in the red blood cells. They collapse in a way that makes the cells look like sickles rather than the round, slightly dished-out shape of functioning red blood cells. The resulting disease is called sickle cell anaemia. It reduces life expectancy in the people in whom the gene is expressed. The cells are incapable of transporting oxygen properly; the disease produces various aches, pains and other forms of debilitation. But how could such a gene prosper? The answer is that sickle cell anaemia confers a relative selective advantage in evolution because the red blood cells of its sufferers are immune to infection by the malaria parasite. Therefore, in areas where malaria is endemic, people with sickle cell anaemia are *relatively* better off and have more surviving offspring than people who contract malaria. But when those offspring were taken as slaves to America, and when a cure for malaria was found, sickle cell anaemia became, again relatively speaking, a pure liability.

The 'racial' link is that the disease is specific to particular populations. When researchers became interested in the disease in the United States, the fashionable tendency was to apply for grants to look into elitist and fancy topics, for example the biochemistry of the sickling process (Michaelson, 1971). This was considered more worthwhile than spending funds for setting up screening and counselling programmes so that potential sufferers could be advised about marriage and having children. Public health, screening, genetic counselling and other activities of this kind in black ghettos are a long way down the pecking order of scientific prestige, and spending time on them is not likely to enhance a scientist's career. However, when black people began to fight for their civil rights, they were able to reorientate funding priorities and to get screening and counselling programmes set up.

Sickle cell anaemia provides a striking example of changing prioritization in research. The search for drugs for the treatment of leprosy or for a simple male contraceptive are examples of other priorities which have been slow to come to the top of the pecking order.

A further possibility – genetic engineering – has emerged as a by-product of other priorities and holds out a long-term hope for a cure for sickle cell anaemia through genetic transplants. Examples of this kind help to show that the real history of science is a series of choices for research which depend, in turn, on matters of class, prestige, gender, and the 'clout' of interest groups. For example, in the same period during which sickle cell anaemia was being ignored, programmes were developed for screening for breast cancer and cervical cancer. Research on blood chemistry which might lead to lower incidence of heart attacks was also well funded. These problems were of great interest to members of the white middle class. It could be argued that most expensive research gets done on diseases that a majority of the world's population does not live long enough to contract. Similarly, many diseases are importantly related to diets which the majority of the world's population has no chance of consuming.

PRIORITIES IN RESEARCH

From those examples one could move on to a whole series of issues about setting priorities in medical research. Approaches to disease through public health measures have been systematically undervalued as compared to approaches which lead to marketable products. Indeed, one of the most striking examples of this concerns the aggressive marketing of a product which is of very little use indeed: powdered milk. The naturally occurring product – mother's breast milk – is more wholesome, contains natural antibodies, and costs nothing.

Racist Society, Racist Science 21

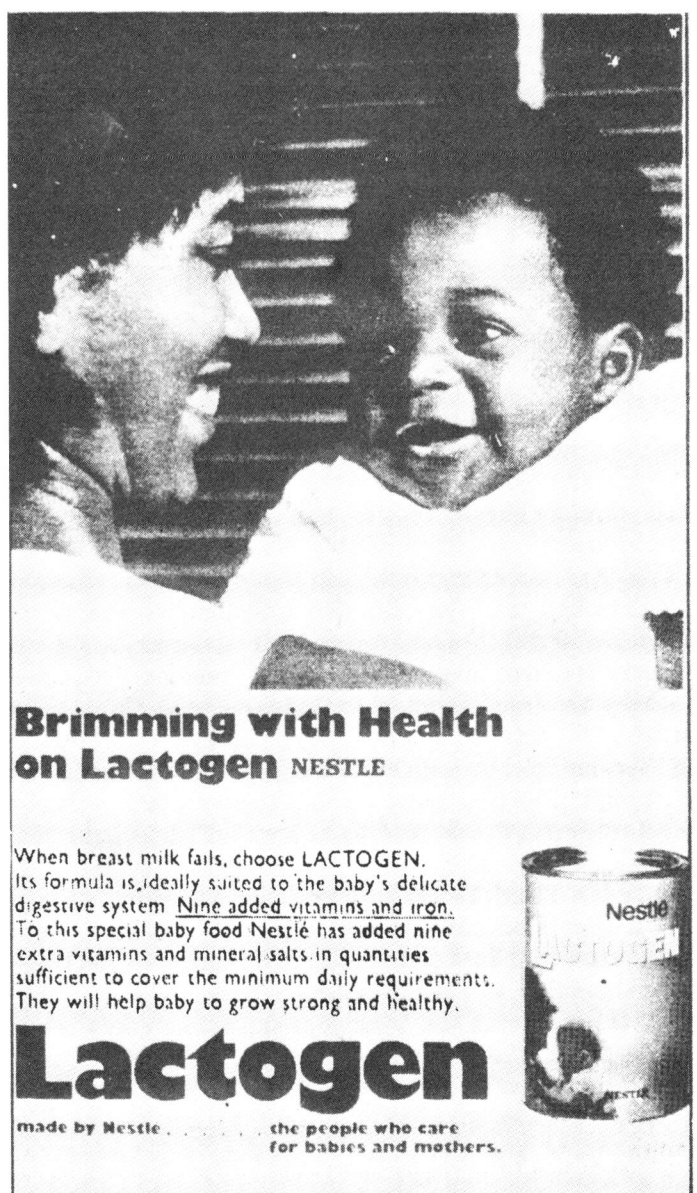

'Nine added vitamins' – better than breast milk?

22 Anti-Racist Science Teaching

This is the limiting case of the creation of marketable products in lieu of other measures. The scandal surrounding the marketing of powdered milk in Third World countries highlights the absurd consequences of commercial priorities.

An anti-racist science curriculum could open out the teaching of science to a historical and social approach to knowledge. This perspective would break down the distinction between the substance and the context of knowledge and examine the social forces and connections (or articulations) of scientific and technological disciplines and research problems. Once one begins to think in this way, many things we already know take on new significance. We also begin to see the blinkering effect of current disciplinary boundaries in the school and university curricula.

The rise of apparently esoteric disciplines makes sense if considered in terms of the power relations in a given society and between communities and power blocs. Remove the conventional barriers and we can see, for example, that the recent dramatic rise in funding for seismology and oceanography (mapping the ocean floor) takes on new meaning in the light of the need to monitor nuclear test ban treaties and to find places to hide nuclear submarines and to find the enemy's. This is not to say that this study of faulting and plate tectonics is wholly explained by these priorities. It only helps us to understand why lavish funds are available for this sort of research.

Indeed, over 40 per cent of scientists in Britain and over 50 per cent of research and development funding comes from the military (Hales, 1982). This is quite often expended on 'pure' science projects which might otherwise not get funded or might not get very much funding. For example, when I was an undergraduate, a professor of marine biology at Yale, Professor Talbot Waterman, got a large grant to go to Bermuda every summer to study crabs who navigated by polarized light in shallow water. This money was given because the United States Office of Naval Research wanted to be able to design ways of flying over the earth's poles, where magnetic compasses do not work.

Similar stories can be told about many seemingly unconnected researches:
☐ Studies of rare-earth metals were orientated towards their use in plane fuselages;
☐ The transport revolution of containerization was a technical spin-off from the development of 'rapid deployment' forces in the United States military;
☐ The development of high-resolution cameras and films was a by-product of spy planes and satellites;
☐ Non-stick frying pans (the example everyone knows) were a by-product of the heat shields used on missile nose cones.

Of course, the money for much of this research comes from the profits of multinational companies which exploit the workers and resources of Third World countries.

It is sometimes hard to grasp the scope of this prioritization process. The whole of the funding of computer-aided design and computer-aided manufacture (CAD/CAM) and numerical control of machines was derived from American military funding (Noble, 1984). The United States armed forces have ambitious plans in this area. General Larry Scance, head of the United States Air Force Manufacturing Command, said the following to a group of contractors (including Westinghouse, Boeing, General Electric):

> Since our war-fighting equipment comes from the industrial base, the condition within that base must be addressed and corrected. We now have an effort under way to provide a planning system that will guide our industrial-based investments and will eventually integrate technology opportunity and business investment planning. It is a top-down approach we call 'industrial base planning'. We plan to maximize application of mechanization and automation, and we plan a paper-free factory with planning, scheduling and control on the latest computer hardware and software techniques. We thus expect the factory that can perform at least one full shift per day unmanned.

Separate out the jargon and you have a fully computerized

24 Anti-Racist Science Teaching

factory without any workers to give you trouble – manufacturing military materials.

The development of cybernetics – the modern science of communication and control – grew out of wartime research on control systems in gunnery and led on to produce new perspectives in a variety of sciences – for example, endocrinology, physiology, psychotherapy, electronics – and connects closely with general systems theory, widely used in management sciences and town planning. (See Haraway, 1981–2; Heims, 1980; Lilienfeld, 1978; Wiener, 1956.)

Nuclear physics is the most obvious and generously funded example where military priorities took a relatively esoteric science and made it into a hugely funded research industry. For example, the vast resources at the European Centre for Nuclear Research (CERN) are a by-product of military priorities but have led to the discovery of new fundamental particles by 'pure' scientists.

Computers from the first generation to the fifth are the result of espionage priorities stemming from World War 2 research to present competition between Japan and the West, with vast resources coming from industry, military, and government (Jones, 1979).

From this need for rapid computation (coupled with developments in planes and missiles) we can derive the whole growth of solid state physics, leading to the transistor, the microprocessor and all the developments extending from pocket calculators and brain scanners to the integrated defence system known as SIOP (Pringle and Arkin, 1983). Someone who is doing research in solid state physics on, for example, selenium arsenide might not be aware of all of the connections of his or her particular PhD project. Indeed, I have a friend doing research in optics at Imperial College who claims that it is extremely hard to avoid doing research which is funded by the military or is of interest to the military. The most abstruse mathematicians have recently found whole areas of their discipline classified.

Even the most advanced and humane research, the transplanting of hearts and other organs, depended on developments in immune system suppression which Sir Peter Medawar and others developed during the treatment of severe military burns during World War 2.

It would be possible to extend this list indefinitely from low-technology matters like modern nursing in relation to the Crimean War, to group psychology and World War 2 stress research, to the entire war-related agenda of the largest private research organization in the world, Bell Labs. From the two-volume history of Bell Labs one can derive an astonishing list of inventions where military and civilian applications were closely integrated (Fagen, 1975, 1978). The same can be said of the history of IBM, the giant which dominates the computer industry.

COMMERCIAL AGENDA-SETTING

A similar story can be told about commercial prioritization and agenda-setting. Vitamins are vital coenzymes; small amounts are necessary to avoid well-known deficiency diseases, for example rickets, pellagra, scurvy. This is a real need at some times and in certain parts of the world, but the vast sales of vitamins in metropolitan countries bears no relationship to the real need. This is simply the result of hype. Yet this same drug industry does not develop cheap vaccines against malaria and other diseases because the potential purchasers of such products cannot afford them (P. Medawar, 1986).

Look at the way the commercial potential of biotechnology has created a bonanza in shares for researchers and genetic engineering and other aspects of the commercial side of biology. They have been frank about their priorities and have gone for products – human insulin, growth hormone – which will lead to expensive saleable commodities (Yoxen, 1982). Indeed, the adding of hormones to animal feeds has already begun to have dire consequences in Puerto Rico, where children are developing secondary sexual characteristics as a result of substances fed to the animals which the children eat.

The United States health industry is operating at over two

THE AMERICA

MORTALITY RATE

1978: BLACK WOMEN die in childbirth at a rate almost four times that of white women. Black women lack prenatal care at a rate twice that of white women. *Children's Defense Fund*, 1980

DISEASE AND DEATH – Minorities compared to White are 3 times as likely to die of hypertension; 4 times as likely to die of kidney problems; 2 times as likely to die of diabetes; and 5 times as likely to die of tuberculosis. *Civil Rights Digest*, Fall, 1977

STATE OF MINORITY HEALTH

INDIAN MALNUTRITION – It should be remembered that disease, malnutrition and obesity were foreign to the aboriginal Indian. Today most elderly Indians are affected by these problems, with malnutrition heading the list. *Insights on Minority Elderly*, 1977

BLACK CHILD MALNUTRITION – 32.7% of all black children suffer from malnutritional deficiencies compared to 14.6% of all white children. *Civil Rights Digest*, Fall, 1977

BLACKS IN STEEL INDUSTRY – Lung and respiratory cancers are high among coke oven workers, where 90% of the workers are black.

JOB RISK: BLACKS are placed in the most dangerous jobs, and face a 37% greater chance of suffering an occupational injury and 20% greater chance of dying from job-related injuries. *Jet Magazine*, 20/12/79

LIFE EXPECTANCY

NAVAJO MEN: Life expectancy for men on the Navajo reservation is 44 years compared with 67 nationally. *New York Times*, 22/4/79

...AY OF DEATH

INFANT MORTALITY

NAVAJOS: The infant mortality rate on the Navajo reservation is 8% compared to a national rate of 2%. *New York Times*, 27/4/79

CHILDREN'S DEATH RATE

The INDIAN SUICIDE rate is twice the national average. On some reservations, it is 5 or 6 times the national average, with rates tending to be highest among young people. *New York Times*, 27/10/80

BLACK AGED

1979: There was 1 black doctor for every 728 black persons, compared to 1 white doctor for every 484 white persons. Children's Defense Fund, 1980

Only 3% of private, government-subsidized housing is inhabited by the black elderly, though 8% of the elderly population is black. National Center on Black Aged, 1977

'American Blacks are facing an ongoing process of genocide. The infant mortality rate for American black babies is higher than fifty-seven per cent of the world's nations. Harlem's 4.3% is typical and is as high as the rate in El Salvador. The ghettos are Third World regions in the midst of a nation teeming with luxury yachts.' **William Bunge, 1983** hundred billion dollars per year. This produces dramatic scientific findings, but their connections should also be spelled out. Much of this money feeds the drug industry, the insurance industry, the medical equipment industry, the hospital construction industry. Beyond this, large profits are derived from the private management of medical facilities and the remuneration of medical practitioners and researchers. Medicine, like other forms of technology, is big business.

The whole question of energy can be seen in the same terms. For example, the debate between various forms of energy derived from the sun – wind, water, heat storage, photovoltaic – is a function of how the big utility companies are able to control research agendas. They have managed to forestall some kinds of research and to co-opt others. The developments which have been done have mimicked the vast power stations that the big utility companies are accustomed to erecting. This is done at the expense of small-scale solar units which can operate in a neighbourhood, a block, even a single house or flat. While this is going on in metropolitan countries, the energy crisis for most of the Third World is more stark. When will the brushwood and charcoal run out? Women forage further and further from home daily in order to find wood with which to cook. Failure to cook food produces disease and debilitation.

Both in the military case and in the setting of commercial priorities, science pursues certain topics and sets up certain research and development agendas at the expense of others. An anti-racist science curriculum could easily move into the detailed examination of these matters and their relations with power politics within and between cultures and power blocs.

CHARITIES: PATRONAGE IN ACTION

Let us look at another aspect of patronage. It would be interesting to tell the whole story of the Rockefeller charities. Before the vast national funding agencies were set up in Britain, Europe and America, the most enlightened and active agent was the Rockefeller charities. Indeed, these provided the model for national funding agencies. Rockefeller wealth was derived from a near-monopoly in the oil industry. The charities were set up for a mixture of altruistic and tax-avoidance reasons. They touch on medicine, public health, social sciences, molecular biology, animal behaviour, the organization of scientific research, First-Third World public health and medical teaching. The history of Rockefeller patronage has been examined by a number of researchers, who have illuminated the following directions:

☐ The United States medical system orientated its professional standards towards the elimination of alternative medical approaches and the setting up of scientifically based medical education, at the expense of a system based on practice and care;

☐ The London School of Hygiene and Tropical Medicine set the pattern for health systems and health practices throughout the Third World;

☐ The Tavistock Institute of Human Relations played a leading part in the use of psychodynamic ideas in industry;

☐ The Yale Institute of Human Relations created a reorientation of all behavioural and social sciences so as to play a palliative role in the American human sciences and in their adoption in other countries;

☐ The whole discipline of molecular biology was virtually created by Rockefeller patronage, which gave funds for the use of physical models and research techniques, e.g. the ultracentrifuge;

☐ The Harvard Business School provided models for management and organization which were deeply influenced by Rockefeller patronage and drew on organic, non-conflict conceptions drawn from physiology;

☐ CIMMYT, the Mexican-based Rockefeller research centre which gave us the Green Revolution, has had the effect of endangering the peasant farmer. This has led to the worldwide spread of high-technology, high-fertilizer farming. It is being followed by a 'new green revolution' using the techniques of biotechnology which are derived from other aspects of Rockefeller patronage;

☐ The monkey colonies in Florida and Puerto Rico provided much of the basis for modern primate research and the models of primate behaviour and family relations which have had a wide influence on the behavioural sciences;
☐ Finally, the style of research funding which was established by the Rockefeller Foundation provided the model for public sector research patronage, the Science Research Council, Medical Research Council, the National Institutes of Health, National Science Foundation.
(See Abir-Am, 1982; Brown, 1979; Dicks, 1970; Fisher, 1978; Fitzgerald, 1986; Fosdick, 1952; Kloppenberg and Kenney, 1984; Kohler, 1976; Morawski, 1986; Yoxen, 1981, 1982).

An analogous story could be told about the history of museums – the custodial homes of scientific progress. Research on the forces which have led to the presentation of knowledge in various museums show how they frame nature and human achievement. There are studies available about research in this vein on the London Science Museum and the New York Museum of Natural History (Levidow and Young, 1984; Stocking, 1985). It is important to understand that the presentation of science to the general public and to schoolchildren plays an important role in the way we think of 'human progress'.

PHILOSOPHY OF SCIENCE

This argument has moved from narrow issues of race and IQ to how agendas are set in science. The above examples have been an attempt to break down the barrier between science and society, between pure enquiry and the sources of prioritization and funding. A way of summarizing the link is to say that a racist society will give you a racist science, in both obvious and unobvious ways. The 'powers that be' in a society will constrain its research and development agenda – military powers, economic forces, the social structure of a society. This is not true only of the funding and of the choice of questions and research proposals. It is also true of ideas of nature and human nature, as well as the philosophical assumptions of science itself. It is harder to make this level of the social constitution of knowledge accessible. One needs to make a big jump.

We are taught to think of science as knowledge of nature pursued by the best method of discovery and proof – the scientific method, which uses quantitative measures of physical variables. But this was not always so. There was science before the 'scientific revolution' in the sixteenth and seventeenth centuries. There are alternative ways of thinking about the world – alternative world-views – in different cultures as well as in our own. If we try to think like anthropologists, we can look at scientists as a tribe and the assumptions of science as a belief system. There is a literature about some of these matters which shows how our 'tribe' and others construct their world-views or cosmologies and set up knowledge systems, technologies and cures within that framework (Douglas, 1970, 1973, 1975; Horton, 1967). The institutions and the educational systems which reproduce them propagate the priorities and values of that tribe.

THE CURRICULUM

A sophisticated anti-racist science curriculum would compare the thought system of our own culture with that of others. It would also show the examples where systems of knowledge have successfully cut across particular cultures – Arabic numbers, acupuncture, herbal remedies. Similarly, it could look at the Western capitalist approach to science – post-seventeenth-century rationalism – and see the concomitant growth of the scientific revolution, the Protestant revolution, and the capitalist revolution. There is a good literature on these matters.

There are both historical and theoretical issues at this fundamental level. For example, the scientific revolution began the separation of fact and value, matter and mind, mechanism and purpose. It would be useful to demonstrate the growth of this mechanical philosophy and compare it to more organismic ways of thinking, then and now. The relationship between mechanistic and reductionist thinking on the one hand, and environmentalist and organismic theories on the other, has important philosophical and political bearings.

It would also be possible to spell out the system which separates facts from their origins, meanings and the values that inhere in knowledge systems, by looking at positivism itself. It is positivistic to teach science in the way we do, such that a science staff in a particular school could make the following reply to a questionnaire on anti-racist science:

1. The nature of Western science is factual and international in character. While clearly the product of European/American historical socioeconomic processes, it purports to be culture free, in that it deals with facts and theories elucidated by a process of logical deduction and reasoning which has its roots in the capacity of the human mind, and not in culturally determined conditions.

2. The dominant form of science in the world today is 'Western' science. The process of logical hypothetical deductive reasoning on which it is based is the subject matter of science education, as is the result of such reasoning expressed in scientific fact and theory.

3. Although many of our pupils originate from divergent cultures, most of them are first-generation Londoners, and their original cultural backgrounds subscribe to Western science. There is therefore no case for introducing other forms of science which are characteristic of particularly minor cultural experiences.

4. The document 'Are We Meeting the Multi-Ethnic Needs of the School?' cannot therefore be replied to by this department, except to state the following:

a) The Science staff are of course aware of the international character of their subjects. This is stressed when appropriate.

b) A 'Statement of Intent' has been produced by the Science Department. This is displayed in the Science Block and every pupil has the opportunity to see it. The statement stresses that staff are actively examining materials to eliminate racial, sexual and religious bias where they are found to occur.

SCHOOLS SHOULD GIVE BALANCED VIEW ON PEACE AND WAR
Sir Keith deplores attempts at indoctrination

The extent to which explicit attention to the issue of peace and war should have a place in the classroom should be decided on educational not political grounds, Education Secretary Sir Keith Joseph said today.

Speaking in London at a one-day conference on peace studies, organized by the National Council of Women of Great Britain, Sir Keith deplored attempts to trivialize the issue, cloud it with inappropriate appeals to emotion and present it one-sidedly.

Arguing for a rational approach to the subject, he said local education authorities should support the professionalism of teachers, who should seek to present to pupils a balanced and objective picture of the issue.

Sir Keith said peace and war, like other important issues of the day, would crop up naturally in the curriculum. There was no need to make special space for studies labelled 'peace'.

Teachers' presentation of the issue should be objective in that their selection of fact gave a balanced picture, fact and opinion was clearly separated and pupils were encouraged to weigh the evidence and arguments so as to arrive at rational judgements.

If asked by his pupils for his own views the teacher should, as appropriate, declare where he himself stands but explain at the same time that others, in particular the pupil's parents and other teachers, may disagree.

Press Release, 3 March 1984

c) Where the subject matter of science includes a discussion of racial origins and differences, as it does in biology, the subject is dealt with in a factual manner.

I think that students should be invited to consider this reply. Similarly, they could be invited to look at Sir Keith Joseph's arguments – as Minister of Education – against the teaching of 'peace' studies on grounds that it amounted to political indoctrination.

The overall model for a science curriculum should be one that always considers *all* the following in their mutual interrelations:
☐ origins
☐ assumptions
☐ articulations
☐ who benefits
☐ alternatives

When we begin to think of alternative perspectives, we can look at science as a way of expressing the values of a given culture. We can also begin to question those who want to maintain the separation between science and culture. For example, Sir Keith Joseph also argued that the arts should be eliminated from all polytechnics in Britain. Why? Students could also be invited to examine the exhibits at various museums. The idea would be to help them to see what lies behind the way scientific knowledge is presented and to see the interests which are being served by the separation of science, technology and society.

The problem of an anti-racist science curriculum is the problem of changing the terrain of science teaching. We need to look at deeply held assumptions about what science is, and what its role in culture is. In particular, there is a series of very basic separations which have to be challenged *in* the curriculum and not just as an ornament to teaching after we have finished presenting the serious part. Some of these separations are:

☐ pure and applied
☐ science and its abuse
☐ science and culture
☐ fact and value
☐ substance and context
☐ body and mind
☐ science and society.

Once teachers and students have become accustomed to thinking about these matters, it should not be difficult to move on to the deeper level of *why* we think about nature in the ways that we do and why we find it so difficult to think about it in different ways. The more we consider these matters, the more closely integrated we will see questions of industry and knowledge, of science and culture, so that in the end debates about the science curriculum can be seen as debates about what kind of society we wish to have. Do we wish to have one in which people are spoon-fed with facts according to an agenda set by others, or one in which they have a genuine voice in determining what our future will consist of? In order to understand that, it becomes necessary to understand the nature of a technocracy – a society based on blinkered technique, while the priorities are set in a way that is kept out of sight. These blinkers are at work in the framing of scientific education. From this set of topics it should be evident that the question of an anti-racist science is the same as the question of a just society in other areas.

References

(All books published in London unless otherwise noted.)

Pnina Abir-Am, 'The Discourse of Physical Power and Biological Knowledge in the 1930s: A Reappraisal of the Rockefeller Foundation's "Policy" in Molecular Biology', *Social Studies of Science* 12 (1982), 341–82.

Pnina Abir-Am, 'How Scientists View Their Heroes: Some Remarks on the Mechanism of Myth Construction', *J. Hist. Biol.* 15 (1982), 281–313.

Ziggi Alexander and Audrey Dewjee, eds, *Wonderful Adventures of*

Mrs Seacole in Many Lands, Bristol, Falling Wall Press, 1984 (nursing, racism, Crimean War).

Martin Barker, *The New Racism: Conservatives and the Ideology of the Tribe*, Junction, 1981.

Peter Berger and Thomas Luckman, *The Social Construction of Reality: A Treatise in the Sociology of Knowledge*, NY, Anchor, 1967.

Howard S. Berliner, *A System of Scientific Medicine: Philanthropic Foundations in the Flexner Era*, Tavistock, 1985.

Morris Berman, *The Re-enchantment of the World*, Cornell University Press, 1981.

B. V. Bowden, *Faster than Thought: A Symposium on Digital Computing Machines*, Pitman, 1953.

Harry Braverman, *Labor and Monopoly Capital: The Degradation of Work in the Twentieth Century*, Monthly Review Press, 1974.

E. Richard Brown, 'He Who Pays the Piper: Foundations, the Medical Profession and Medical Education', in S. Reverby and D. Rosner, eds, *Health Care in America: Essays in Social History*, Philadelphia, Temple University Press, 1979, pp. 132–54 (Rockefeller).

E. Richard Brown, *Rockefeller Medicine Men: Medicine and Capitalism in America*, University of California Press, 1979.

Edwin A. Burtt, *The Metaphysical Foundations of Modern Physical Science*, 2nd edn., Routledge, 1932.

Arthur L. Caplan, ed., *The Sociobiology Debate*, Harper & Row, 1978.

Allan Chase, *The Legacy of Malthus: The Social Costs of the New Scientific Racism*, University of Illinois Press, 1980.

Anthony Cheetham, ed., *Science Against Man*, Sphere, 1971.

Andy Chetley, *The Baby Killer Scandal*, War on Want, 1979.

Stephan L. Chorover, *From Genesis to Genocide: The Meaning of Human Nature and the Power of Behavior Control*, MIT Press, 1980.

Harry Cleaver, *Origins of the Green Revolution*, PhD dissertation, Stanford University, 1975.

Robin Cook, *Coma*, Pan, 1978.

Roger Cooter, *The Cultural Meaning of Popular Science: Phrenology and the Organization of Consent in Nineteenth-Century Britain*, Cambridge University Press, 1984.

Ruth Schwartz Cowan, *More Work for Mother: The Ironies of Household Technology from the Open Hearth to the Microwave*, NY, Basic, 1983.

C. D. Darlington, *The Evolution of Man and Society*, Allen & Unwin, 1969.

C. D. Darlington, *The Little Universe of Man*, Allen & Unwin, 1978.

H. V. Dicks, *Fifty Years of the Tavistock Clinic*, Routledge & Kegan Paul, 1970.

David Dickson, *Alternative Technology and the Politics of Technical Change*, Fontana, 1974.

David Dickson, *The New Politics of Science*, NY, Pantheon, 1984.

Mary Douglas, *Purity and Danger: An Analysis of Concepts of Pollution and Taboo*, Penguin, 1970.

Mary Douglas, ed., *Rules and Meanings: The Anthropology of Everyday Knowledge*, Penguin Education, 1973.

Mary Douglas, *Implicit Meanings: Essays in Anthropology*, Routledge, 1975.

Albert Einstein, 'Why Socialism?', in L. Levidow, ed., *Radical Science Essays*, pp. 214–20.

M. Fagen, ed., *A History of Engineering and Science in the Bell System*, 2 vols, Bell Telephone Laboratories, 1975, 1978.

Karl Figlio, 'Sinister Medicine? A Critique of Left Approaches to Medicine', *Radical Science Journal* 9 (1979), 14–68.

Karl Figlio, 'Medical Diagnosis, Class Dynamics, Social Stability', in Levidow and Young, eds, *Science, Technology and the Labour Process*, vol. 2, pp. 129–65.

D. Fisher, 'The Rockefeller Foundation and the Development of Scientific Medicine in Great Britain', *Minerva* 16 (1978), 20–41.

Deborah Fitzgerald, 'Exporting American Agriculture: The Rockefeller Foundation in Mexico 1943–53', *Social Studies of Science* 16 (1986), 45–83.

Raymond B. Fosdick, *The Story of the Rockefeller Foundation*, NY, Harper & Row, 1952.

Carl Gardner and Robert M. Young, 'Science on TV: a Critique', in T. Bennett *et al.*, eds, *Popular Television and Film: a Reader*, BFI/Open University Press, 1981, pp. 171–93.

Susan George, *How the Other Half Dies: The Real Reasons for World Hunger*, Penguin, 1976.

Rachel Grossman, 'Women's Place in the Integrated Circuit', *Pacific Research*, 9, 5–6 (1978), 2–17.

Mike Hales, *Science or Society? The Politics of the Work of Scientists*, Pan, 1982; Free Association Books, 1986.

Donna J. Haraway, 'The High Cost of Information in Post World War II: Evolutionary Biology: Ergonomics, Semiotics, and the Sociology

of Communications Systems', *The Philosophical Forum*, 13, 2–3 (1981–2), 244–78.
David Harvey, 'Population, Resources, and the Ideology of Science', *Economic Geography* 50 (1974), 256–77.
L. S. Hearnshaw, *Cyril Burt: Psychologist*, Hodder & Stoughton, 1979.
Steve J. Heims, *John von Neumann and Norbert Wiener: from Mathematics to the Technologies of Life and Death*, MIT Press, 1980.
David Himmelstein and Steffie Woolhander, eds, *Science, Technology, and Capitalism*, Special Issue of *Monthly Review*, 38, 3 (1986).
Andrew Hodges, *Alan Turing: The Enigma*, Burnett Books, 1983.
Luke Hodgkin, 'Mathematics as Ideology and Politics', in L. Levidow, ed., *Radical Science Essays*, pp. 198–213.
Donald Horne, *The Great Museum: The Re-Presentation of History*, Pluto Press, 1984.
Robin A. Horton, 'African Traditional Thought and Western Science', *Africa* 37 (1967), Part I: 'From Tradition to Science', pp. 50–71; Part II: 'The Closed and Open Predicaments', pp. 155–87.
R. V. Jones, *Most Secret War: British Scientific Intelligence, 1939–1945*, Coronet, 1979.
L. J. Jordanova, 'Natural Facts: A Historical Perspective on Science and Sexuality', in C. MacCormack and M. Strathern, eds, *Nature, Culture and Gender*, Cambridge University Press, 1980, pp. 42–69.
T. J. Kalikow, 'Konrad Lorenz's "Brown Past": A Reply to Alec Nisbett', *J. Hist. Behav. Sci.* 14 (1978), 173–80.
Leon J. Kamin, *The Science and Politics of IQ* (1974), Penguin, 1977.
Martin Kennedy, *Biotechnology: The University-Industrial Complex*, Yale University Press, 1986.
Daniel J. Kevles, *In the Name of Eugenics: Genetics and the Uses of Human Heredity*, NY, Knopf, 1985.
Philip Kitcher, *Vaulting Ambition: Sociobiology and the Quest for Human Nature*, MIT Press, 1985.
Jack Kloppenburg and Martin Kenney, 'Biotechnology, Seeds and the Restructuring of Agriculture', *Insurgent Sociologist*, 12, 3 (1984), 3–17.
Karin D. Knorr-Cetina, 'The Constructivist Programme in the Sociology of Science: Retreats or Advances?', *Social Studies of Science* 12 (1982), 320–328.

Karin D. Knorr-Cetina and Michael Mulkay, eds, *Science Observed: Perspectives on the Social Study of Science*, Sage, 1983; critique by Tim Rowse, 'Sociology Pulls Its Punches', in L. Levidow, ed., *Science as Politics/Radical Science* 20, Free Association Books, 1986, pp. 139–49.

R. E. Kohler, 'The Management of Science: The Experience of Warren Weaver and the Rockefeller Foundation Programme in Molecular Biology', *Minerva* 14 (1976), 279–306.

R. E. Kohler, 'A Policy for the Advancement of Science: The Rockefeller Foundation, 1924–29', *Minerva* 16 (1978), 480–515.

Leszek Kolakowski, *Positivist Philosophy from Hume to the Vienna Circle*, Penguin, 1972.

Joel Kovel, *Against the State of Nuclear Terror*, Pan, 1983; Free Association Books, 1986.

B. Latour and S. Woolgar, *Laboratory Life: The Social Construction of Scientific Facts*, Sage, 1979.

Les Levidow, 'A Marxist Critique of the IQ Debate', *Radical Science Journal* 6/7 (1977), 13–72.

Les Levidow, ed., *Radical Science Essays*, Free Association Books, 1986.

Les Levidow and Bob Young, 'Exhibiting Nuclear Power: The Science Museum Cover-Up', in Radical Science Collective, eds, *No Clear Reason: Nuclear Power Politics/Radical Science* 14, Free Association Books, 1984, pp. 53–79.

Les Levidow and Bob Young, eds, *Science, Technology and the Labour Process: Marxist Studies*, 2 vols, vol. 1, CSE Books, 1981; Free Association Books, 1983; vol. 2, Free Association Books, 1985.

Richard Levins and Richard Lewontin, *The Dialectical Biologist*, Harvard University Press, 1985; commentary by Peter Taylor, 'Dialectical Biology as Political Practice', in L. Levidow, ed., *Science as Politics/Radical Science* 20, Free Association Books, 1986, pp. 81–111.

George Lichtheim, *The Concept of Ideology and Other Essays*, NY, Vintage, 1967.

Robert Lilienfeld, *The Rise of Systems Theory: An Ideological Analysis*, NY, Wiley, 1978.

Donald MacKenzie and Judy Wajcman, eds, *The Social Shaping of Technology*, Milton Keynes, Open University Press, 1985.

Karl Mannheim, *Ideology and Utopia: An Introduction to the Sociology of Knowledge*, Routledge, 1960.

40 Anti-Racist Science Teaching

Herbert Marcuse, *One-Dimensional Man: The Ideology of Industrial Society*, Sphere, 1968.
Armand Mattelart, 'Infotech and the Third World', in Radical Science Collective, eds, *Making Waves: The Politics of Communication/ Radical Science* 16, Free Association Books, 1985, pp. 27–35.
Charles Medawar, *Insult or Injury? An Enquiry into the Marketing and Advertising of British Food and Drug Products in the Third World*, Social Audit, 1979.
Peter Medawar, *Memoir of a Thinking Radish: An Autobiography*, Oxford, 1986.
Carolyn Merchant, *The Death of Nature: Women, Ecology and the Scientific Revolution*, Wildwood House, 1982.
Michael Michaelson, 'Sickle Cell Anemia: An "Interesting Pathology"' *Ramparts* (Oct. 1971), 52–8, reprinted in this collection.
Mary Midgley, *Evolution as a Religion: Strange Hopes and Stranger Fears*, Methuen, 1985.
Pat Roy Mooney, *Seeds of the Earth: A Private or Public Resource?*, International Coalition for Development Action, 1979.
Pat Roy Mooney, *The Law of the Seed: Another Development and Plant Genetic Resources*, special issue of *Development Dialogue*, 1983: nos 1–2.
J. G. Morawski, 'Organizing Knowledge and Behavior at Yale's Institute of Human Relations', *Isis* 77 (1986), 219–42.
Mike Muller, *The Health of Nations: A North–South Investigation*, Faber & Faber, 1982.
Greg Myers, 'Texts as Knowledge Claims: The Social Construction of Two Biology Articles', *Social Studies of Science* 15 (1985), 593–630.
Norman Myers, ed., *The Gaia Atlas of Planet Management*, Pan, 1985.
David F. Noble, *Forces of Production: A Social History of Industrial Automation*, NY, Knopf, 1984.
Peter Pringle and William Askin, *STOP: Nuclear War from the Inside*, Sphere, 1983.
Radical Science Journal Collective, 'Science, Technology, Medicine and the Socialist Movement', *Radical Science Journal* 11 (1981), 3–70.
Ken Richardson and David Spears, eds, *Race, Culture and Intelligence*, Penguin, 1972.
Jeremy Rifkin, *Algeny*, NY, Viking Press, 1983.

Janet Sayers, *Biological Politics: Feminist and Anti-Feminist Perspectives*, Tavistock, 1982.
George W. Stocking, Jr, 'Philanthropoids and Vanishing Cultures: Rockefeller Funding and the End of the Museum Era', in Stocking, ed., *Objects and Others: Essays on Museums and Material Culture*, History of Anthropology, vol. 3, Madison, University of Wisconsin Press, 1985, pp. 112–45.
Jon Turney, ed., *Sci-Tech Report: Current Issues in Science and Technology*, Pluto Press, 1984.
P. G. Werskey, *The Visible College*, Allen Lane, 1978.
Alfred North Whitehead, *Science and the Modern World*, with an Introduction by R. M. Young, Free Association Books, 1985.
Norbert Wiener, *The Human Use of Human Beings: Cybernetics and Society*, NY, Anchor, 1956.
Cecil Woodham-Smith, *Florence Nightingale, 1820–1910*, McGraw-Hill, 1951.
Donald Worster, *Nature's Economy: A History of Ecological Ideas*, Cambridge University Press, 1985.
Robert M. Young, 'Evolutionary Biology and Ideology: Then and Now', *Science Studies* 1 (1971), 177–206, revised in W. Fuller, ed., *The Biological Revolution*, NY, Anchor, 1972, pp. 241–82.
Robert M. Young, 'The Human Limits of Nature', in J. Benthall, ed., *The Limits of Human Nature*, Allen Lane, 1973, pp. 235–74.
Robert M. Young, 'Science *is* Social Relations', *Radical Science Journal* 5 (1977), 65–129.
Robert M. Young, 'Interpreting the Production of Science', *New Scientist* (29 March 1979), 1026–8, reprinted in this collection.
Robert M. Young, 'Science as Culture', *Quarto* 2 (Dec. 1979), 8.
Robert M. Young, 'Science is a Labour Process', *Science for People* 43 (1979), 31–7.
Robert M. Young, 'Why are Figures So Significant? The Role and the Critique of Quantification', in J. Irvine and I. Miles, eds, *Demystifying Social Statistics*, Pluto, 1979, pp. 63–75.
Robert M. Young, 'The Naturalization of Value Systems in the Human Sciences', in *Problems in the Biological and Human Sciences*, Block VI of Open University Course on Science and Belief from Darwin to Einstein, Milton Keynes, Open University Press, 1981, pp. 63–110.
Robert M. Young, 'Darwinism *is* Social', in D. Kohn, ed., *The Darwinian Heritage*, Princeton, Princeton University Press, 1985, pp. 609–38.

Robert M. Young, *Darwin's Metaphor: Nature's Place in Victorian Culture*, Cambridge University Press, 1985.

Robert M. Young, 'Is Nature a Labour Process?', in Levidow and Young, eds, *Science, Technology and the Labour Process*, vol. 2, pp. 206–32.

Robert M. Young, 'The Dense Medium: Television as Technology', *Political Papers* 13 (1986), 3–5.

Edward Yoxen, 'Life as a Productive Force: Capitalising upon Research in Molecular Biology', in Levidow and Young, eds, *Science, Technology and the Labour Process*, vol. 1, pp. 66–122.

Edward Yoxen, 'Constructing Genetic Diseases', in P. Wright and A. Treacher, eds, *The Problem of Medical Knowledge: Examining the Social Construction of Medicine*, Edinburgh, Edinburgh University Press, 1982, pp. 144–61.

Edward Yoxen, *The Gene Business: Who Should Control Biotechnology?* Pan, 1983; Free Association Books, 1986.

RACISM IN SCIENTIFIC INNOVATION

LES LEVIDOW

How can science be racist? After all, science is supposed to be objective, defined as keeping facts separate from values, and so finding out about reality in a manner free from subjective bias.

Of course, there are a few notorious examples of scientists who have developed theories on the inferiority of black people. For example, Arthur Jensen and Hans Eysenck have made claims for innate racial differences in intelligence, partly drawing upon papers published by Sir Cyril Burt. The revelation that Burt faked his data has made it easier for critics to dismiss such racist claims as merely bad science or pseudo-science. That kind of dismissal serves to exempt the rest of science from social criticism, as if the problem were a few racist apples in the barrel, or 'ideology' polluting an otherwise neutral science.

This essay will suggest just the opposite: how science more generally perpetuates racism. Rarely is this done in conscious, explicit, obvious ways. Nor is it done simply by concocting scientific justifications for the existing order.

Rather, scientific innovations provide more effective ways of exploiting labour and natural resources, as well as ways of undermining people's resistance. In a world where black and Third World people face the most intense exploitation, science and technology invariably have racist effects. Those effects are not simply abuses or misapplications of science, since they arise from racist priorities in defining what counts as the problem for science to solve.

To illustrate those racist priorities, this essay will briefly survey some innovations – in population control, agribusiness, energy and microelectronics – as a guide to further study and reading. The examples are interconnected at many levels. At the most general level, they all serve the international division of labour, by which multinational corporations attempt to manipulate entire populations. To maximize profitability they must continually attempt to control, and alter, where and how people live and work, if they live and work at all.

Early capitalism used sheer physical coercion, starting with slavery and indentured servitude, to force people to produce commodities for a mass market, while modern capitalism uses wage-labour instead (Rowling, 1987). In much of the Third World, where agribusiness has increasingly taken over the best land to produce cash crops for export, people are formally free to sell their labour but are no longer free to produce the means of their existence. Through the threat of poverty and even starvation, millions of people are induced to make themselves available for wage-labour, however and wherever the multi-national corporations want them, as human components of a 'global assembly line'. Such firms have established multiple sources for various stages of production and can transfer production from one country to another; such flexibility reduces workers' bargaining power everywhere (CIGE, 1984).

In that context, population control does not arise in response to 'overpopulation'; nor do new agricultural technologies respond to food shortages, any more than energy policies respond to some 'energy gap'. Rather, all such innovations serve the multinationals' quest for global power over labour and resources.

POPULATION CONTROL

The explosive growth of the human population is the most significant terrestrial event of the past million millennia. Three and one-half billion people now inhabit the Earth, and every year this number increases by 70 million. Armed with weapons as diverse as thermonuclear bombs and DDT this mass of humanity now threatens to destroy most of the life on the planet. Mankind itself may stand on the brink of extinction; in its death throes it could take with it most of the other passengers of Spaceship Earth. No geological event in a billion years – not the emergence of mighty mountain ranges, nor the submergence of entire subcontinents, nor the occurrence of periodical glacial ages – has posed a threat to terrestrial life comparable to that of human overpopulation. (Paul Ehrlich, *Population, Resources, Environment*, 1972)

The above quotation, written by a leading scientist and best-selling author, comes from a textbook widely used in university courses on the environment. While Ehrlich rightly sees nuclear weapons and pesticides as threats to humanity, he speaks as if it were the masses of people who were armed with such weapons. In reality, of course, the weapons are borne by governments and multinational corporations, which deny the people any real decision-making over those threats to humanity. They also collude to deny people the fruits of their own labour and – if profitability demands it – to deny people the means to produce anything at all. Furthermore, as Barry Commoner (1980) has

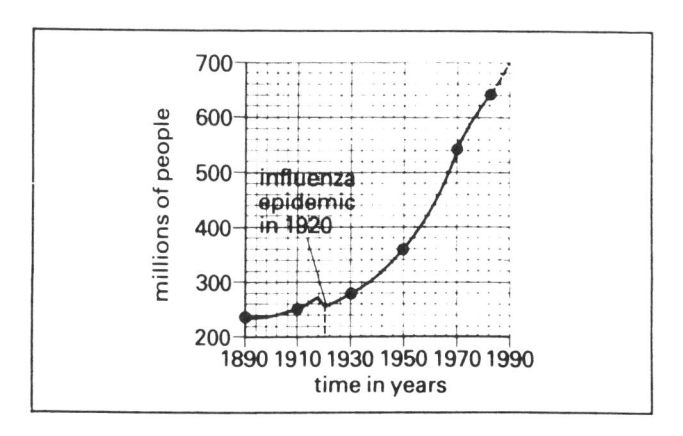

School textbooks typically identify the problem as 'overpopulation' caused by lack of birth control, while ignoring the reasons why so many Third World people are denied the means to buy or produce food.

argued, 'overpopulation' itself is a misleading term because it attributes poverty to large family size rather than the other way around, while obscuring the real causes of poverty.

Having started with a provocative quotation from a leading scientist on population control, this section will focus on the institutional framework for scientists' work in that area. When they develop new means of sterilization or contraception, try them out and assess the results, they do so with certain definitions of the problem to be solved and certain criteria for success. Where do these come from? On a world scale the US multinationals, and the governments who serve them, have provided much of the research brief and resources. In particular the US Agency for International Development (AID), the Rockefeller Foundation and the International Planned Parenthood Federation (IPPF) have all played leading roles.

What are these organizations' interests in population control? Listen to Alan F. Guttmacher, a recent President of the IPPF:

> Reckless population growth without parallel economic growth ... makes for a constant lowering of the standard of living. Such a decline, with its concomitant mounting poverty and hunger, inevitably delivers a population to some kind of -ism, whether it be communism, fascism, or Pan-Arabism, and weans them away from democracy.

Although Guttmacher's warning remains vague about the relationship between population growth and economic growth, he clearly echoes Ehrlich in attributing poverty to 'reckless' breeding, which he apparently sees as analogous to frivolous joy-riding. More importantly, the IPPF's priority is clear: not to empower Third World people but to keep them loyal to 'democracy', by one means or another. As we will see, in practice this has meant using population control to overcome obstacles to exploiting such people. Let us examine some examples, starting with the IPPF's pursuit of 'democracy' in Puerto Rico. (The following sketch is based on *SftP*, 1973, 1974; Mass, 1976; film, *Le Operacion*.)

In the 1930s, US corporations were using Puerto Rico to

produce abundant food – not for local people, but for export, with some re-imported from the USA in processed form at high prices. The result was conveniently called 'overpopulation' rather than a planned shortage of cheap, locally grown food. The USA thereby justified reducing the colony's potentially rebellious population through a sterilization programme. For employers it had the added advantage of freeing more women from child-bearing, instead to labour in sewing shops. Since then a local affiliate of the IPPF has sterilized fully one-third of Puerto Rican women, often with the pretence that the operation was reversible.

Furthermore, as part of a wider strategy of depopulating the island, its women have been used as the prime guinea pigs for testing new contraceptive methods. Contraceptives with dangerous side effects were tested at the Humacao Clinic for three years before the dangers were revealed. Present-day contraceptive foam was first tested on Puerto Rican women – at twenty(!) times its present strength – without, of course, informing the women that the chemicals were previously untested. At the same time, the US Agency for International Development (AID) increased its budget for birth control while reducing its budget for health care by a similar amount. Employers in Puerto Rico gladly gave their women workers time off work to visit the birth control clinic in the factory – thereby saving themselves the expense of maternity benefits.

Puerto Rico is only the most extreme example of a more widespread US policy, both at home and abroad. It is no surprise that American doctors recommend sterilization far more frequently for their state welfare patients than for private patients, and especially for mothers of illegitimate children; nor is it surprising that nearly half of all women sterilized under federally financed programmes have been black, especially as state funding for abortions has declined. Increasingly, Puerto Rican immigrants and black women in US cities have found themselves subjected to sterilization, often without their consent or even knowledge; sometimes this is achieved by imposing emotional manipulation on the women just before they give birth. Among the native Amerindian population, roughly 40 per cent of the women of child-bearing age in the 1970s were sterilized in a similar way, many by the US government-controlled 'Indian Health Service'. In reducing their birth rate, one aim was apparently to weaken the people's resistance to corporations keen to mine the uranium, coal and other minerals on Indian land.

In the USA these racist practices have been institutionalized and even legitimized through quasi-medical terms. Naturally, the so-called 'habitual aborter' deserves sterilization. The euphemism 'Mississippi appendectomy' has come to describe the 'hysterilization' (hysterectomy) commonly inflicted upon indigent black women. The term 'Puerto Rican Syndrome', used to describe those people's emotive protests against degrading treatment, actually found its way into an American paediatrics textbook describing suicidal adolescents! (Rodriguez-Trias, 1978.)

Throughout the world – El Salvador, Colombia, Bangladesh, India – AID has promoted sterilization, compulsory in all but name. In one of the more recently exposed scandals, Bangladeshi women were denied food aid unless they agreed to be sterilized (BIAG, 1985). Furthermore, Third World women are systematically used as guinea pigs. In New York, Puerto Rican women have been unknowingly used for psychological experiments on enduring childbirth 'with and without emotional support' from those present. An AID front organization, Pathfinder Fund, has operated all over Latin America to distribute contraceptives and do IUD experiments. And AID has helped to dump unsafe contraceptives – such as the Dalkon Shield and Depo Provera – on the Third World after they have been found unsafe in the USA (Mother Jones, 1979).

More recently, AID has set up programmes to train Third World doctors in techniques for reversing sterilization and for researching types of sterilization easier to reverse. The aim: to make sterilization more attractive to women by portraying the operation as fully reversible, even though such facilities will be available at most to a few urban women.

Lest these horror stories appear far removed from basic research, it is worth recalling that the neo-Malthusian mentality justifying population control has been embedded both in laboratory experiments and in the Verhulst-Pearl equations describing a supposedly natural law of population growth. Who funded such research? Again, none other than the Rockefeller Foundation (Clarke, 1984, pp. 156–7).

All these examples are intended to illustrate how apparent 'abuses of science' are integrally connected to research priorities on the one hand and to multinational exploitation on the other.

AGRIBUSINESS

Multinational corporations have tied Third World economies into a global system that impoverishes and underdevelops its people. In that context, sterilization programmes have coincided with political decisions to oppose any reforms benefiting the people. Furthermore, they form part of a wider strategy:
☐ slowing down or blocking land reform;
☐ increasing investment in agribusiness to increase productivity of cash crops for export;
☐ devising counterinsurgency schemes involving the military in social and economic programmes; and
☐ promoting 'family planning' among populations who refuse to work at low wages in profitably productive ways (Cleaver, 1977).

In the West, racist panics over 'the population explosion' have attributed Third World poverty, and even wars, to its people's supposed fecklessness or inadequacies. Such images have provided legitimation for imperialism to reimpose its control in the name of saving the world from overpopulation and instability.

Integral to its strategy was the Green Revolution, based on new high-yield varieties (HYVs) of grain. As in the IPPF outrages described earlier, here too the Rockefeller Foundation initiated the genetic research – in Mexico, where the government had nationalized the Rockefeller family's land holdings. Through HYVs the multinationals could regain their profits by controlling the conditions of production, without themselves needing to own the land. The HYVs turned out to help to intensify farmers' dependency upon agribusiness, since the high yields required regular purchase of the seeds, pesticides, fertilizer and equipment. Intensified market competition impoverished the already-poor small peasants, turned them into landless labourers, and made them less able to buy the new product. So the higher yields meant greater exports of wheat or rice, less production of high-protein crops, and deforestation of vast land tracts. Meanwhile, the US government used food as a political weapon in many ways, such as withholding emergency shipments to famine-struck India until that country agreed to adopt the HYVs (Hales, 1982).

In the USA itself the struggle of Chicano farm workers has been world-renowned for a long time, but less well known is the role of science and technology in the story. The expansion of the California tomato industry began in the 1940s with the development of the Pearson tomato, a high-yielding variety specially suited to that state's growing conditions, with the added condition of pesticides. To overcome the wartime labour shortage while avoiding trade-union militancy, the 'bracero programme' was devised to import 'cheap', unorganized labour from Mexico. The Pearson tomato – and its racist exploitation of labour – became widespread as the programme continued after the war. By the late 1960s – to counter the strength of the United Farm Workers – growers had started mechanizing the harvest with the help of new tomato varieties; government-funded scientists had specially bred them for uniform size, uniform ripening and rough handling by machines.

Furthermore, the migration of the braceros into the Rio Grande valley depressed wages there and led Chicano workers – US citizens of Mexican descent – to migrate to the midwestern states, where they brought their traditions of union organizing. Finally, by the 1970s, the employers there too were introducing mechanized harvesters and related changes in food processing in order to reduce their dependence on rebellious workers (Kitron, 1984).

This story illustrates how innovations are geared to the racist and class exploitation of Third World people. It arises from a definite choice of research problems to be solved – in this case researched with government funds at a Land Grant College. Of course, an equivalent effort could have designed instead new varieties for easier picking by farm workers and for pest control without expensive pesticides. Such criteria have had no place among the priorities of those who make funding decisions, serving the food processors' interests.

The vanguard of biological research, biotechnology, may appear benign as compared to the examples already given here. But its agricultural developments, dubbed 'the second Green Revolution', seem headed to worsen the Third World's dependence on agribusiness firms. Designed to obtain more output per unit investment, the new techniques will make it even more difficult for small farmers to survive. These techniques involve the multinationals designing new plant varieties from gene stocks taken from the Third World, patenting the seeds, and then selling them back expensively to countries whose populations suffer from malnutrition (Yoxen, 1983).

Genetic engineering techniques, a basis of biotechnology, are also being directed towards those medical applications predicted to be the most lucrative for the health industry. Enormous sums are spent to develop medicines affordable by relatively few people, mostly in Western countries. Meanwhile, with malaria affecting 150 million people and threatening up to two billion in the Third World, research teams have been blocked from developing anti-malarial vaccines because of disputes between the World Health Organization and Genentech over public access versus private patenting of research results. Another disease, sickle cell anaemia, has been known for a long time to affect many people of West Indian, African and Mediterranean origin, yet few resources have been made available for the genetic screening, counselling and diagnosis services that could alleviate the disease's harmful effects and save lives; and there is little sign of any serious commitment of genetic engineering resources, which could potentially develop a genetic transplant to prevent the disease.

ENERGY

A paradox: amidst the so-called 'energy crisis', both biotechnology and automation consume large amounts of energy in order to reduce employers' dependence on paid labour – thereby creating an even greater surplus of human energy, or unemployment. At the same time, for most of the world's population, the 'energy crisis' has little to do with oil or coal, since it is experienced as a shortage of firewood, a shortage aggravated by the sorts of imperialist 'development' already described here. In some places firewood has become such a commercial item – available mainly to the rich – that the poor rely increasingly upon cow dung and agricultural wastes for their fuel source (Agarwal, 1982). Meanwhile, potentially cheap, small-scale alternatives – solar power or biogas plants – are being developed on large-scale models, inaccessible to the people who most need them.

Of course, the main priority for energy research continues to be nuclear power, geared to supplying electricity for urban dwellers and industry. This has racist effects, not only in the

"Ma, what is heavy water?"

global context of fuel usage and dependence on the West, but also in more specific ways. In some Western countries, immigrants are employed as casual labourers – for example, North Africans in France – to perform the tasks that most expose workers to radioactivity. In the routine language of the industry, these workers are 'burnt out' to the maximum permissible limit, then replaced by others (Jungk, 1979). In India such tasks are performed by illiterate villagers whose cancer deaths years later are attributed to fate or misfortune. Even among educated professionals in India, their lack of any real control over radioactivity dangers has led to a resurgence of traditional religion and superstition – an ironic result of Western-induced dependence upon dangerous imported technology (Sharma, 1984).

MICROELECTRONICS

Microelectronics, enthusiastically promoted for its 'labour-saving devices', seems unlikely to save labour for the working class. Indeed, it provides a basis for new, improved ways of exploiting Third World people, both in the Western metropolis and abroad.

First and foremost, underlying the entire microelectronics industry is the mass production of integrated circuits etched on silicon chips, whose most labour-intensive stage is based in South East Asia. Young women's health is ruined as they inhale noxious fumes, and their eyesight is ruined by peering through microscopes (Grossman, 1978). Similar work is done by Asian immigrants in Silicon Valley, California.

All this forms part of a 'global assembly line', both in the production of microelectronics and in their application (Fuentes and Ehrenreich, 1983). The ease of transferring work around the world helps multinationals to exploit workers more effectively on a global scale. By threatening to transfer work from workers who resist low wages and poor conditions, such firms weaken potential opposition.

For decades now, labour-intensive processes have been transferred from imperialist to dominated countries in search

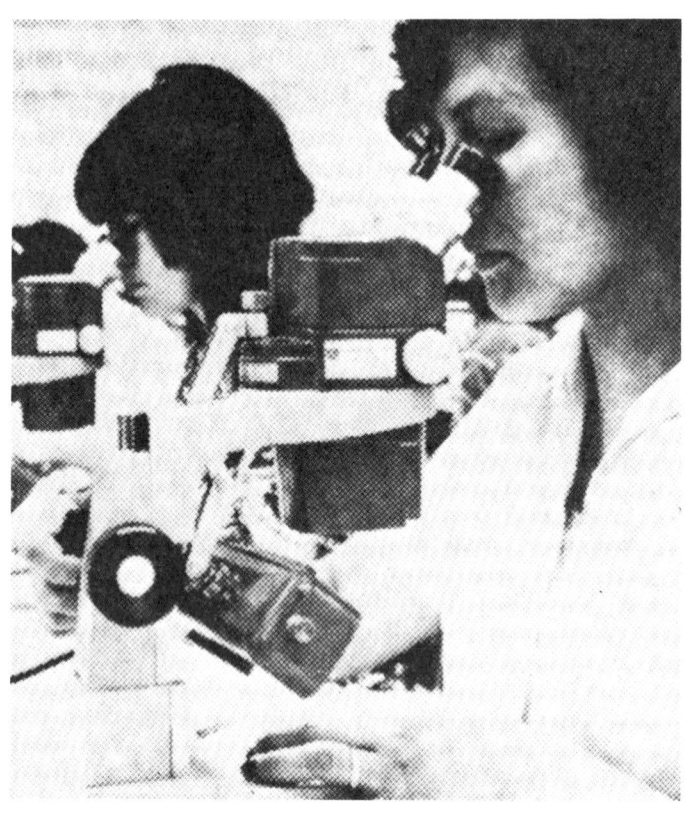

Expendable lungs and eyesight . . . 'cost-efficient labour', as advertised by the Malaysian government

of cheap labour. Now the cheapening of automated systems may shift the trend back the other way. In any case, the widened options give the multinationals another bargaining weapon against workers' resistance the world over.

Automation can have specifically racist effects in the metropolitan countries as well. Consider the example of the Grunwick photo processing plant in London, which drew nationwide attention during the 1976–7 strike there. The firm's expensive automation of the chemical processing and customer accounts made economic sense only as the basis for more intensive, racist exploitation of the Asian women in the mail order department. Not only did management keep the basic wages low, but it used threats and humiliation to compel the workers to accept overtime at short notice, in order to keep up with peak flows of mail order work.

Although that control temporarily broke down with the strike, the trade unions' paralysis allowed the firm's management to win. Thus it vindicated the new model for capitalism's future – a model since extended to the nationalized industries. While Grunwick was commonly seen as a vestige of the nineteenth-century sweatshop, it actually stands as capital's vanguard of twentieth-century automation, which creates anew the basis for sweatshop conditions (Levidow, 1981; *Year of the Beaver*, 1985).

That example also illustrates the emerging two-tier polarization among those who succeed in staying in paid work despite automation. Inevitably, preparation for semi-employment or menial jobs becomes a significant orientation of secondary schooling, likewise of the ever-growing Manpower Services Commission programmes such as the Youth Training Scheme – YTS, popularly renamed 'You Tried Slavery?' (Cohen, 1984). These are intended especially for school leavers lacking formal qualifications, which means in particular black youth. Regardless of whether such youths join or reject the schemes in the face of DHSS threats to withdraw social security benefits, the work discipline that they impose must be understood as an integral part of the 'microelectronics revolution'.

With the inevitable unemployment and underemployment, there comes an essential concomitant: the new technology of repression, imported from active duty in Northern Ireland. These include not only riot technologies – plastic bullets and water cannon – but also computers to collate information collected from many sources, including electronic surveillance (BSSRS, 1985). A key information source continues to be informers, often unwitting accomplices to the police strategy of 'targeting and surveillance'; the targets include those categories of people whom police prejudices brand as criminal suspects, especially black youth. Through 'multi-agency policing', police gain access to student records and classrooms, as well as informal access to teachers' familiarity with students – all of which serves the purpose of targeting such people (ACPS, 1986).

Thus an apparently benign 'public service', multi-agency-policing, connects to computers, at least in two ways. First, it directly expands police files, held on national and local computers. Second, it helps the police to target those youths it considers potentially subversive – that is, those likely to protest the limited options on offer from a high-tech economy imposing unemployment and underemployment. This kind of policing is perhaps the ultimate weapon against those who refuse to accept the more stringent conditions set for selling their labour and buying back their lives with wages.

WHOSE SCIENCE?

In conclusion, these examples are intended to show how a capitalist society produces a racist science. They illustrate how racist priorities influence the kinds of science that actually get done, and therefore the kinds of effects that necessarily follow.

In combating the 'Bhopal Syndrome', must science be irrevocably part of the problem? Or could it be part of the solution? Is a different kind of science possible? Perhaps, but only if we abandon the pretence of a value-free knowledge. That means formulating research goals from the standpoint of how oppressed people define their own interests and needs, rather than from the standpoint of profitability and capitalist work discipline.

In that regard, the predominant science education is racist, at the very least by omission. It fails to acknowledge the racist effects of science, much less its hidden values and priorities. Instead it portrays science as mainly basic disinterested research, not responsible for its applications. That aspect – teaching future scientists not to see the oppressive connections – has been an integral part of their training. Like science itself, science teaching cannot be neutral, apolitical or non-racist. Either it hides these connections or reveals them; it is either part of the problem or part of the solution.

Suggested Reading

(All books published in London unless otherwise noted.)

ACPS, *Policing Schools*, 1986 (PO Box 447, London E2 9PS).
Anil Agarwal, 'Firewood: Fuel of the Rich', *Science for People* 52 (Summer 1982), 12–13.
Grazyna Baran *et al.*, 'GASP! A Critique', *Science News* 28 (Nov. 1984), 22–3, reprinted in this collection.
BIAG, *Food, Saris and Sterilization*, 1985, £2 from Bangladesh International Action Group, PO Box 94, London N5 1UN.
Lynda Birke and Jonathan Silvertown, eds, *More Than the Parts: Biology and Politics*, Pluto, 1984, pp. 110–35.
BSSRS Technology of Political Control Group, *TechnoCop: New Police Technologies*, Free Association Books, 1985.
Richard Clarke, 'Population, Poverty and Politics', in Birke and Silvertown, *op. cit.*, pp. 152–76.
CIGE, 1, 3 (1984): *The Global Economy: Trade, Aid and Multinationals* (Contemporary Issues in Geography and Education).
Harry Cleaver, 'Food, Famine and the International Crisis', *Zerowork* 2 (1977), 7–70.
Philip Cohen, 'Against the New Vocationalism', in I. Bates *et al.*, *Schooling for the Dole? The New Vocationalism*, Macmillan, 1984, pp. 104–69.
Barry Commoner, 'How Poverty Breeds Overpopulation', in Arditti *et al.*, eds, *Science and Liberation*, Boston, South End Press, 1980, pp. 76–90.
Paul Ehrlich, *Population, Resources, Environment: Issues in Human Ecology*, Freeman, 1972.
Dan Finn, 'Leaving School and Growing Up: Work Experience in the Juvenile Labour Market', in Inge Bates *et al.*, *Schooling for the Dole? The New Vocationalism*, Macmillan, 1984, pp. 17–64.
Dan Finn, *Training Without Jobs*, Macmillan, 1986.
Anette Fuentes and Barbara Ehrenreich, *Women in the Global Factory*, Boston, South End Press, 1983.
Rachel Grossman *et al.*, *Changing Role of S.E. Asian Women*, 1978, available at $2 from PSC, 222B View Street, Mountain View, CA 94041, USA.
Mike Hales, *Science or Society?*, Pan, 1982; Free Association Books, 1986, pp. 36–45.
Robert Jungk, *The Nuclear State*, John Calder, 1979.

58 Anti-Racist Science Teaching

Uriel Kitron *et al.*, 'The Tomato is Red: Agriculture and Political Action', in Birke and Silvertown, *op. cit.*, pp. 177–95.

Les Levidow, 'Grunwick: The Social Contract Meets the 20th Century Sweatshop', in L. Levidow and B. Young, eds, *Science, Technology and the Labour Process: Marxist Studies*, vol. 1, CSE Books, 1981, reissued by Free Association Books, 1983, pp. 123–71.

Bonnie Mass, *Population Target: The Political Economy of Population Control in Latin America*, Ontario, Latin American Working Group, 1976.

Mother Jones reprint, 'The Corporate Crime of the Century', Nov. 1979, available from 625 Third Street, San Francisco, CA 94107.

Le Operacion, film available from The Other Cinema, London.

Dr Helen Rodriguez-Trias, 'Women and the Health Care System', 'Sterilization Abuse', 1978, available at $1 from WIRE, 2700 Broadway, NY, NY 10025.

Nick Rowling and Phil Evans, *Commodities: How the World Was Taken to Market*, Free Association Books, 1987.

SEAD, *Electronics and Development: Scotland and Malaysia in the International Electronics Industry*, Edinburgh, 1984 (available from SEAD, 29 Nicolson Square, Edinburgh EH8 9 BX).

SftP, V, 2, (Mar./Apr. 1973) and VI, 1 (Jan./Feb. 1974), available from Science for the People, 897 Main St, Cambridge, MA 02139.

SftP, XVII, 3 (May/June 1985): Decoding Biotechnology.

Dhirendra Sharma, 'India's Nuclear Estate: An Interview', in *No Clear Reason: Nuclear Power Politics / Radical Science 14*, Free Association Books, 1984, pp. 94–106.

Lenny Siegal and John Markoff, *The High Cost of High Tech: The Dark Side of the Chip*, Bessie / Harper & Row, 1986.

Nancy Stepan, *The Idea of Race in Science: Great Britain, 1800–1960*, Macmillan, 1982.

Year of the Beaver, Faction Films, 1985 (available from The Other Cinema, London).

Edward Yoxen, *The Gene Business*, Pan, 1983; Free Association Books, 1986, especially Chapter 5.

SICKLE CELL ANAEMIA: AN 'INTERESTING PATHOLOGY'

MICHAEL G. MICHAELSON

I

Invariably the patient is black, invariably young, perhaps your own age, in fact, a thought which may disturb you when they bring her out on stage for you to see, for you to question. She (or he) has been wheeled from her hospital bed for your convenience, your medical education. It is a lecture in biochemistry, most likely, and you have been reading about blood and about haemoglobin, the molecule red blood cells contain; the protein with a backbone of iron, the carrier of whatever pure oxygen can be sucked from polluted air. By now you know haemoglobin well, you have memorized it: its four poly-peptide chains, two alpha, two beta (the 'globin' part of the molecule); its four iron-containing 'haem' groups; its foetal and adult forms; its electrophoretic patterns; its abnormal varieties: haemoglobin C and M and E and – this is the powerful one, the terrible one – haemoglobin S. One of its black victims is in front of you now, probably frightened, sitting before strange, mostly white, mostly male, faces. She is in a wheelchair, looking up for a moment, meeting your curious stares, then looking down at the floor.

'Tell these young doctors something about yourself, Miss Williams,' the instructor says, smiling. 'Tell them why you are in the hospital.'

This is an exercise in 'clinical correlation', an attempt on the part of the medical school to demonstrate that the 'basic science' courses of your first year or two are not irrelevant or as sterile as they seem. This is what the catalogue calls 'contact' with patients, 'early exposure' to 'clinical material'. You have been looking forward to it, bored by years of organic chemistry and physics, lectures and diagrams. You are nervous, perhaps, like Miss Williams, but you are proud of your knowledge, and confident, like a doctor. You observe her closely: her arms and legs are long and skinny and gangling; her abdomen is short but full and protruding; her legs are heavily bandaged; her eyes are yellow; intravenous fluids run into her arm; she is dressed in a limp hospital nightgown; she is weak and difficult to hear.

'I have sickle cell disease,' she says quietly. 'Sickle cell anaemia.' Then she tells you her story, which typically goes something like this:

She is twenty-four years old and has been in the hospital eighteen times. She would get tired easily ever since she was a child; she did not develop as rapidly or as well as the other children; she would often get colds and more severe respiratory infections. At age seven her tonsils were removed. At age eight she was hospitalized for a month with severe joint pain, fever, and heart murmurs, diagnosed incorrectly as rheumatic fever. At twelve she developed ulcers on both ankles which would not heal, even now, twelve years later. She then began having periodic attacks of severe, incapacitating pain in her bones and joints, her back and her abdomen. These 'crises' would last about a week, leaving her exhausted, weaker than ever, and sore all over.

When she was fourteen her appendix was removed. At sixteen her spleen was removed. In the last eight years she has had six more hospital admissions for the treatment of her recurrent leg ulcers, which have eaten down to the layer of muscle and bone. Two admissions for pneumonia, two for long episodes of fever, chills, night-sweats and diarrhoea. Now, she tells you, she is urinating blood, and her belly is swelling like a balloon and her heart, her doctors have told her, is too big and is getting tired. She cannot sleep lying down, but only on three pillows, and she wakes up at night gasping for breath. Her eyes have been yellow for six years and her liver, she knows, is not working right. In her life she has received (if she remembers right) eighty-six transfusions of whole blood.

'Thank you, Miss Williams,' the instructor says, and you awkwardly mumble thanks too, as she is wheeled back to the hospital.

In the brief discussion that follows, you learn that there is no cure for sickle cell anaemia, nothing, in fact, that substantially helps a patient endure a crisis. You learn that very few patients with the disease – which strikes one Black in five hundred – manage to survive to the age of thirty. You guess, as you leave the lecture room with friends for lunch, that Miss Williams, with the soft voice and the memories of pain and the present of pain, will be dead before you practise medicine. But in fact, she is dead much sooner than that.

II

These days sickle cell anaemia has a certain *chic* in academic medical circles. The disease has always been considered an 'interesting pathology', but never, until very recently, has it been considered very *important*. The distinction is real and its consequences disturbing.

Sickle cell anaemia is one of the very few fatal diseases which is known (that is, thought) to be caused by a demonstrable inherited abnormality – an identifiable, highly specific, biochemical aberration in the molecular structure of haemoglobin. As a consequence, it has for two decades been waved by medical scientists before the noses of their occasional critics: evidence that modern medical science is figuring things out, making progress. But if the disease has provided some comfort for chemists and first-year medical students, it has been relentless with its victims, the hundreds of thousands of Blacks who, as they suffer and die, give their bodies as subjects for the scientists and exhibits for the students. Although the disease is found in one in five hundred black babies, and although these childen survive only an average of twenty years, and although very much more is understood of the genetics, chemistry and epidemiology of sickle cell than of other serious diseases of childhood, physicians and researchers within the white establishment have virtually ignored it. Their own children are safe.

62 Anti-Racist Science Teaching

Right now in the United States some fifty thousand Blacks are dying of SCA; nearly two million have 'sickle cell trait', a milder form of the disease which usually (but not always) leaves them symptom-free but which enables them to pass the trait or the disease on to their children. 'In 1967,' according to an article published last year [1970] in the *Journal* of the American Medical Association, 'there were an estimated 1,155 new cases of SCA, 1,206 of cystic fibrosis, 813 of muscular dystrophy, and 350 of phenylketonuria. Yet volunteer organizations raised $1.9 million for cystic fibrosis, $7.9 million for muscular dystrophy, but less than $100,000 for SCA.' In fact, the figure for sickle cell anaemia was closer to $50,000.

There was no nationwide organization devoted to sickle cell anaemia until last year; no celebrity has ever done a sickle cell telethon or chaired a committee. The Research Grants Index of the National Institutes for Health for fiscal year 1968 indicated that grants for much less common childhood illnesses greatly exceeded those for SCA. In one year research into cystic fibrosis, for example, received 65 NIH grants, although this disorder is found in only one in three thousand births (98 per cent of cystic fibrosis victims, it is worth noting, are white). The same year there were 41 grants for phenylketonuria affecting one in ten thousand (again, all of them Whites), a disease which can be detected and controlled by diet, and which is not fatal. Yet there were fewer than two dozen grants for research into sickle cell anaemia. A review of the facts, even the editors of the conservative AMA *Journal* were forced to conclude, 'clearly focuses a spotlight on public failure to recognize the importance of combating sickle cell anaemia . . .'

The *Journal* had found it 'incredible . . . that very few people in the black population at large have been offered pertinent information about sickle cell anaemia and the mode of its transmission.' But the Black Panthers, for instance, do not find it incredible; they find it, on the contrary, to be expected and have launched a major drive through their People's Free Clinics to provide the black community with information about the disease along with free diagnostic blood tests. Black Panther leaflets distributed in Philadelphia and other cities, in an attempt to provide 'pertinent information' about SCA and the times and places for blood tests, are headlined in large block letters above pictures of normal and sickled red cells: BLACK GENOCIDE.

Perhaps because of the notoriety, doctors are finally beginning to take an interest; the research on SCA is picking up and money is starting to flow. In his 'health message' of a few months ago Richard Nixon announced that he would request six million dollars in congressional funds to 'combat' sickle cell anaemia. A deputy director of the National Heart and Lung Institute was named on 12 May as co-ordinator of a new federal sickle cell disease programme. Physicians are suddenly competing with one another in attempts to invent and patent new and better and cheaper screening tests and effective therapeutic procedures. School systems, the United States Army, pharmaceutical houses and medical technology companies are entering the field.

But Dr Roland B. Scott, a black physician working on sickle cell research for twenty years, a generation before it became fashionable, has never received a federal penny for his efforts. Now chairman of the department of paediatrics and director of the Sickle Cell Center at Howard University, he has offered this evaluation of the current boom: 'All is chaos now, with a lot of researchers running around flag-waving, headline-grabbing, trying to qualify for the therapy and screening jackpot, with the government looking for quick results for high visibility in 1972. But it'll fade away, and we Blacks will be left with the problem.'

Almost certainly he is correct. The current enthusiasm for sickle cell anaemia may in the end reveal less about this killing disease than about contemporary American medical politics. It may be part of the larger effort to 'save' not black children but an obsolete and elitist system of medical care which has oppressed patients of all races and classes (though especially women, Third World people, and the poor, of course) for a century and which is now, at last, on the verge of collapse. And if past experience is an index, the current fad can be expected, as Dr Scott suggests, to come and go, leaving the black people of this country in their usual condition of ill-health and powerlessness, while providing their (largely white) professional and corporate 'benefactors' with new avenues for advancement and prestige, profits and power.

III

Any issue, however isolated, which even for a moment appears to bring Richard Nixon and the Black Panther Party to the same side of an argument surely merits careful analysis. In the case of sickle cell anaemia, the discussion seems to require at least a brief review of the disease state itself, its biochemistry, genetics, and medical 'history'. And this must be placed in the context of the now much-publicized 'crisis' in medical care, the rise in the last ten years of a liberal medical elite and, more recently, the emergence of a broadly based insurgent radical health movement.

The distinguishing characteristics of sickle cell anaemia were first described in the medical literature in 1910 by James B. Herrick, a prominent mid-western cardiologist. He wrote of a twenty-year-old West Indian student whose symptoms involved 'a secondary anaemia . . . strikingly atypical in the large number of nucleated red cells of the normoblastic [immature] type and in the tendency of the erythrocytes [red blood cells] to assume a slender, sickle-like shape.' Herrick was perplexed by this 'most bizarre group of symptoms', and he could offer no definitive diagnosis, although he suspected syphilis.

The case accumulated. Dr Emmel (1917) was the first to use the term 'sickle cells' repeatedly; Mason (1922) the first to refer to 'sickle cell anaemia'; Graham and McCarthy (1926) the first to point out that this anomaly of red cells afflicted only Blacks. In 1927 and 1930 further studies suggested that the sickling of red cells occurs when the haemoglobin they contain passes from the oxygen-saturated state to the unsaturated (reduced) state. The abnormal haemoglobin would apparently alter the shape of these red cells, distorting their normal disc-like, biconcave 'doughnut' configuration and producing, somehow, the characteristic physical symptoms. The disease was labelled a 'haemoglobinopathy' and passed on, in effect, to the biochemists.

In 1949 Linus Pauling reported that normal haemoglobin (A) and haemoglobin derived from sickle cell patients (S) demonstrated different patterns of 'electrophoretic mobility', molecular movement in an electrical field. Ingram (1961) perfected the electrophoretic technique and was able to show that haemoglobin S differs from A only in the replacement of a single amino acid (glutamate) in the alpha chain of the normal molecule by another (valine) in the abnormal molecule. It is this single amino acid in haemoglobin's total 287 which is wrong; this one which exacts so high a price.

The original difficulty, however, is genetic: every child receives from his parents two genes which determine the kind of haemoglobin the child's body will synthesize. Only haemoglobin S will sickle when deprived of oxygen, and only the black population (less than two per cent of SCA's victims are non-black, usually of Mediterranean origin) carries this particular abnormal gene in its 'genetic pool'. If both parental genes are for haemoglobin S (the homozygous state, SS) the child will suffer from the disease. If only one gene is abnormal (heterozygous, AS), the child will have the trait and will be able to transmit the S haemoglobin to his own children, although he will not, most likely, be fatally ill himself.

Under certain circumstances these heterozygotes were well off. It has been suggested, and is generally agreed, that sickle cell trait confers upon its bearers a protection against malignant forms of malaria. On the African continent, particularly where malaria is endemic, the AS haemoglobin type represented a protective evolutionary adaptation. When heterozygotes married, they would, according to Mendelian genetics, bear children one quarter of whom would inherit two 'normal' genes for haemoglobin (AA), and most of these would die of malaria; another quarter would be homozygous for sickle cell (SS) and would die of that disease; one half, statistically, would be heterozygous (AS), without the disease, but bearing the trait, and therefore protected against malaria. The neat evolutionary scheme, however, made no provisions for slave ships.

In the United States resistance against malaria is not a significant advantage, since malaria is not a significant threat. But evidence is beginning to mount that sickle cell trait, which affects nearly two million Blacks in this country, is less harmless than was formerly believed. Last year the AMA *Journal* and *Lancet*, a British medical journal, published reports of bleeding episodes and sudden death in patients undergoing 'routine' surgery, due apparently to respiratory depression during surgical anaesthesia. The *Journal* of the National Medical Association published a similar report more recently. An editorial in *The New England Journal of Medicine* on 14 May 1970 noted that although the trait has 'generally been regarded as a benign condition', there have been 'occasional reports of splenic, pulmonary, pituitary or cerebral infarction, as well as priapism, haematuria, hyposthenuria, avascular necrosis of bone and retinal haemorrhage', indicating 'that in certain unfavourable circumstances these patients may experience the severe vaso-occlusive episodes characteristic of homozygous sickle cell anaemia.' The 'unfavourable circumstances' included surgical anaesthesia, aeroplane flights, athletic and military training and severe infections – all conditions involving increased bodily demands for oxygen, decreased environmental oxygen tension, or both. Most dramatic were these four case reports of black recruits 'undergoing army basic training at a post at an altitude of 4,060 feet. All were apparently healthy and had no family history of anaemia . . .'

> CASE 1. W. J., a 21-year-old man, collapsed and lost consciousness after a 40-yard low crawl and a 300-metre run during his first day of training. After admission to the hospital he regained consciousness and complained of shortness of breath and faintness. On examination he was acutely ill . . . One hour later he became hypotensive and combative and lost consciousness. Despite vigorous treatment for hyperkalemia [an electrolyte imbalance] and hypotension, he remained comatose. Nasal haemorrhage

increased. He died 24 hours after the initial collapse. At autopsy, the lungs were heavy, and serosanguineous [thick bloody] fluid flowed from the cut surfaces. The bowel was filled with dark bloody fluid. The vessels in all microscopical sections were packed with sickled red blood cells ... Haemoglobin electrophoresis revealed SA haemoglobin.

CASE 2. M. T., a 19-year-old soldier, collapsed while running a mile during his twenty-first day of training. Upon arrival at the medical clinic no pulse or spontaneous respirations were present. Resuscitation was successful ... Complete unresponsiveness was the only positive physical finding ... Bright red blood was oozing from the gastric aspirate; blood oozed from venous puncture sites, and a lower-gastrointestinal-tract haemorrhage developed ... There was no urine output. Bleeding continued; hypotension [low blood pressure] followed, and despite volume and fluid replacement, the patient died 25 hours after he collapsed ... Haemoglobin SA was found on electrophoresis.

CASE 3. V. H., a 21-year-old man, complained of faintness after a 20-yard crawl during his first day of training and suddenly lost consciousness. He was dead on arrival at the medical clinic ... Histologic sections of all organs displayed massive congestion of the vasculature, with sickling of virtually all red blood cells. Haemoglobin electrophoresis of post-mortem blood revealed SA haemoglobin.

CASE 4. L. T., a 21-year-old recruit, complained of faintness and numbness of the legs and lost consciousness while running once around his barracks after arriving for training. On admission he had regained consciousness and complained of pain and weakness in the legs. He was in no acute distress ... General physical and neurological examinations were unremarkable but he suddenly became apneic [ceased breathing] ... He died eight hours after collapsing ... A sickle cell preparation at autopsy was positive.

The army was unimpressed by the reports. Although a number of physicians suggested that Blacks with sickle cell trait be draft-exempt, the military still does not require even diagnostic screening tests of new recruits. A Fort Dix pathologist wrote to the *New England Journal* to complain that evidence supporting a relationship between sickle cell trait and the sudden death of army recruits was 'insufficient', and a Defense Department spokesman has insisted that 'personnel found to have the trait are still considered qualified for general duty.' While army spokesmen suggest that coincidence or drug abuse might have been implicated in the reported deaths, the Black Panther Party Newspaper suggests that 'the racist US power structure has no intention of ceasing this form of genocide.' On 22 May 1971 the paper announced: 'Therefore the Black Panther Party is initiating a programme to help research really begin that can eventually discover the cure and prevention of Sickle Cell Anaemia.'

IV

The motivation and the aim of the Panther programme seem unimpeachable. For years sickle cell patients have been paraded before beginning medical students as evidence that biochemistry is relevant, that genetics can be fun. Yet for all this 'understanding', no treatment or effective palliation exists for the disease, and no real effort has been made to increase public awareness of its nature: if recent samples are representative, less than one-third of the black population of this country has ever heard of it. Finally, one black person in ten carries a gene for haemoglobin S, unaware of that fact and of its possible consequences.

But in the light of what is known of sickle cell anaemia – and, more important, of what is not known – serious questions arise. Granted the obvious truth that black people, like all people, have a right to all information which relates to their health and well-being, how is such information likely to be used? Blacks with sickle cell trait should of course know that if they bear children with another heterozygote there is a 25 per cent chance that those children will suffer from sickle cell disease and die. But what are the limits of 'genetic counselling?' Where does advice and information end and compulsion begin? Perhaps Blacks with the trait should be exempted from military service. But should they be excluded from professional sports? From riding in aeroplanes? Will their insurance rates be raised as, for example, the premiums of people with even mild diabetes are raised? (Here it should be emphasized that sudden death in sickle cell trait remains an exceedingly rare phenomenon; heterozygotes have a normal life expectancy and are not invalids.) As a health worker at the Mark Clark People's Free Health Center in Philadelphia I have participated in the Panther screening programme, and I hope to continue to do so. But there have been painful moments when I have wondered if, in telling someone in otherwise perfect health that he or she had 'sickle cell trait', we were not forcing that person to pay a price in anxiety and confusion disproportionate to what she had gained. Would she get sick? Probably not. Could we help her? No. What should she do? Nothing; except avoid marryng someone else with the trait; or have no children; or adopt children; or have children at her own risk; that is, at theirs.

More troubling, I think, is the Party's statement that its programme will 'help research really begin that can eventually discover the cure and prevention' of SCA. Such a goal is unquestionably admirable. But the level of hope and expectation it reflects, as is the case with most claims made of and by 'modern medical science' (whether for wonder drugs or organ transplantation), seems to be dangerously inflated. Although, on the one hand, a cure for sickle cell anaemia might appear to be close because its specific 'cause' has been identified, that same fact reasonably may, on the other hand, give rise to a sobering scepticism. Precisely because so much is known about the disease and has been known for some time, the lack of therapeutic progress thus far is especially discouraging.

According to the elegant, even seductive, theory of sickle cell pathophysiology, the various but typical symptoms of the disease are related to the unique property of haemoglobin S: under conditions of low oxygen concentration the molecule forms crystals called 'tactoids' which stretch the cells to their abnormal elongated shape, predisposing to circulatory obstruction in small vessels where blood cells pass one at a time. This sickling phenomenon occurs more easily with higher percentages of S haemoglobin (i.e. SS rather than AS), in acidic solutions, and under conditions of stasis (slow or stopped blood flow). Thus tissues which normally contain relatively acidic blood which is relatively low in oxygen (e.g. kidneys, lungs) are particularly susceptible to this vicious circle: sickled blood cells become trapped in small capillaries, leading to stasis, deoxygenation, increased acidity, and increased sickling; the sequence is repeated until blood clots, disrupted (occluded) circulation, pain, swelling, haemorrhage, and tissue death occur.

What initiates such a crisis remains (significantly) unknown; infection, stress, high altitudes have been implicated. But since sickling is a reversible phenomenon, at least in test tubes, it has seemed reasonable to believe that a crisis could be relieved by the administration of high-concentration oxygen, anticoagulants (to break up clots), whole blood transfusions, antacids, and so on. These have all been repeatedly tried; and they have all miserably, and significantly, failed. Recent trials with urea (a chemical which returns sickled cells to normal shape in test tubes) have seriously endangered some patients and have been implicated in the death of at least one child.

Even if sickling could be reversed in a patient's circulation, it is impossible to predict whether or not the disease would thereby be 'cured'. Juvenile-onset diabetes, for example, is another disease characterized by a specific chemical difficulty, an inability of the body to secrete insulin. Until this was discovered and insulin made available on a mass scale, its victims died in childhood. Now they live longer, but eventually succumb to degenerative changes of the nervous and circulatory systems which fatally persist despite insulin therapy. The biochemistry has been 'corrected', but the disease state progresses, apparently unimpressed. Even if sickled cells could be 'unsickled', then, victims of SCA might still be in trouble.

Finally, there seems to be the assumption underlying the Panther programme that if more money and more data and research subjects were provided, a cure would somehow follow. Contradicting this are not only the rather depressing results in the case of SCA, outlined above, but the fact that those childhood diseases which *have* received money more extensively – cystic fibrosis, muscular dystrophy, and many others – are as mysterious as ever, as incurable. Then heart disease, stroke, and cancer – well-known scourges of the white middle and upper classes, which have been funded astronomically relative to SCA – might be mentioned as examples of more spectacular failures.

This is not to argue that research into sickle cell anaemia should not be energetically pursued, or that education and screening programmes should not be carried out. As an organizing and fund-raising tool, the People's Fight Against Sickle Cell Anaemia is of unquestionable value; and the history of the white medical establishment's neglect of SCA could hardly be more educational. But the new emphasis on sickle cell screening should not be allowed to deflect energies from other screening programmes for lead poisoning and anaemia (both of which more commonly afflict black people than SCA and which can most often be successfully treated) and from efforts to provide free, dignified 'primary' medical care, Pap tests, prenatal care, and so on.

The liberal medical establishment's new enthusiasm for this 'interesting pathology' of Blacks should not be allowed to obscure the fact that it has done so little, and is still doing little to prevent the malnutrition which causes anaemia, the peeling lead paint which poisons babies, the rats and the roaches. And when money for SCA begins to appear – if in fact it ever does (reportedly Nixon has earmarked the federal research funds for white friends in the South, part of the Southern strategy for 1971) – it should be regarded with a certain caution. Because as Nixon and the doctors enter the sickle cell sweepstakes, armed with programmes and patents and promises, a new set of dangers presents itself: that black people will continue to be used, as they have traditionally been, as 'research and teaching material' by white doctors; that whatever research does take place will be designed by and for physicians and will have as its primary aim professional advancement in academic medical circles rather than increasing the likelihood (however slim) of direct therapeutic benefit to the masses of black people; that 'genetic counselling' from outside the black community might evolve into social control or economic sanction.

V

The most far-reaching consequences of the sickle cell controversy seem to me not 'medical' in the strictest sense, but social and political. It is no accident that the current explosion of interest in sickle cell anaemia is taking place in the midst of the most significant turbulence in the history of American medicine.

Today the failure of this country's 'health care system' is a widely acknowledged, abundantly documented, and massively publicized fact. The American Medical Association, remarkable even in this country for its history of self-interest and exploitation, is beginning at last to disintegrate. Public dissatisfaction with doctors and hospitals is rising, especially (and encouragingly) among the middle classes.

As the crisis intensifies, medical image-makers have begun to sense that the status and economic privileges of physicians in this country can be saved only by rapid legerdemain. Galled by the AMA's failure to take a more 'liberal' stance on matters of medical politics – group practice, health care for the poor, national health insurance – medical-school-based physicians have begun to take the initiative. Always somewhat alienated from community practitioners, whom they consider rather dull and unscientific, these academic medical men, armed with liberal rhetoric, corporation consultants, cost-benefit economics and computers, are manoeuvring for control of medical practice.

The pattern is not without precedent. What is happening today in American medicine happened in almost every other sphere in the 1930s: promises of a 'new deal' are being made in an elegant attempt to dissipate energies for genuinely radical alternatives. Medicine's lag of forty years, however, has not been detrimental for the movement, but providential. For in that time period the deficiencies of corporate liberalism – centralized bureaucracy wholly unaccountable to the public, a distorted sense of priority which places profits and power before service to people, institutionalized depersonalization bordering on mass schizophrenia – have become evident enough. And as AMA conservatives and the urban academic medical elite fight it out, the discontent swells.

It is a perfect moment for the radical health movement – a coalition of consumers, community people, health workers and 'professionals' dedicated to providing free medical care to all people in their local communities, ending economic and status distinctions among health care, calling for community control of all health institutions (including the new liberal medical empires). On 15 June the Medical Committee for Human Rights launched a National Health Crusade which has been gathering a remarkable momentum; it is based upon these principles:

1. End profit-making in health care. Health care is a service, not a business.
2. Pay for all services with a progressive tax on total wealth. One without loopholes that makes corporations and the rich pay their share.
3. Provide complete and preventive care with no charges for health services.
4. Administer medical centres locally through representatives of patients and health workers.
5. Create a federal non-profit corporation to produce and distribute drugs and medical supplies.

People's Free Medical Centres, similar to those of the Black Panther Party, the Young Lords, and Young Patriots and other community groups (there are now an estimated 200 functioning or in preparation throughout the country) are an integral part of this health movement's efforts to revolutionize 'medical care' not merely in the narrow sense but as part of a larger effort at restructuring and making 'healthy' a social order

Counselling session in Brent, at one of Britain's few sickle cell centres

which is conspicuously unwell. These health centres are 'counter-cultural' in the best sense, providing radical alternatives to existing medical (and social) institutions and ideologies. They constitute an active focus for community organizing, education, co-operative work, the formation of new kinds of relationships, communication, sharing, play, love. Some day they may include day-care centres, recreational facilities, places for old people to live and work and be and feel useful. They are, I believe, precious revolutionary resources.

The liberal medical elite, for its part, will fight to subvert this programme and prevent community-worker control of its own institutions; it will try to head off any counter-institutional development which threatens to disrupt the white-male-professional-dominated hierarchy of medical care in this country. And it is in this context that the sudden discovery by the medical establishment and Richard Nixon of sickle cell anaemia becomes rather more clear. Their recent infatuation with SCA is an attempt to prove that American doctors *are* responsive to the needs of the black community; it is an attempt to deflate pressures for radical change, and perhaps the beginning of an attempt to gain control of People's Medical Centres; it is, as Dr Scott has suggested, although not in so many words, a rip-off.

I am not suggesting that government and/or medical empire money and equipment (if and when it becomes available for SCA) should not be accepted. It should be accepted, with a knowing smile, with both hands out for more, indeed with militant demands for more. For the current liberal enthusiasm about sickle cell anaemia leaves deliberately unexamined the real roots of the black community's chronic ill-health. It does not get at poverty or vastly inadequate nutrition, at racism or peeling lead paint, capitalism or rats, at the absence of easy access to free, dignified medical care, at professionalism or sexism, at what a doctor 'is', indeed at what a human being is, and can be, and must be. The health movement *is* beginning to get at those things, and shall continue to as it grows and struggles to serve the people, as the workers at the Mark Clark Center say, body and soul.

Editors' note:
For the situation in Britain see the book by Usha Prashar, Elizabeth Anionwu and Milica Brozovic, *Sickle Cell Anaemia: Who Cares?*, Runnymede Trust, 1985.

INTERPRETING THE PRODUCTION OF SCIENCE

ROBERT M. YOUNG

It is a fact about the division of labour among branches of knowledge that working scientists do not find it any easier to think in sophisticated ways about the social relations of science than sociologists do to think about, say, quantum mechanics or catastrophe theory. But it is also a fact – about the pecking order of who has a right to feel more complacent than whom – that working *natural* scientists tend to think that their niche in the division of labour has a wider field of vision than, say, sociologists – even (especially?) sociologists of science. My own experience is that someone who does research on some aspect of 'science and society' is much less likely to accept an invitation to speak on superconductors, nerve-muscle physiology, X-ray crystallography or psychology than experts in those scientific sub-disciplines are to write and lecture on epistemology, medieval science and society, science and values or the social relations of scientific knowledge.

In some ways this is lovely, because it shows some attempt on the part of specialists to take seriously broader issues of social responsibility. On the other hand, the pronouncements of working scientists – especially eminent ones – tend to be made with an authoritativeness that is not appropriate outside their own area of expertise. It is easy to forget that equally serious people have conducted laborious researches and thought very hard about many aspects of the social relations of science, technology and medicine, and that disciplines, sub-disciplines and attendant literatures have grown up which are worth looking into before pronouncing quite so 'authoritatively'.

I say all this because I have recently had a number of experiences of working scientists writing in ways which I have found pretty dismissive and dogmatic about various aspects of science and values, science and society, science and ideology, in current and historical terms. These writings have betrayed little or no awareness that there are ongoing debates on these matters among people who have done their homework. The reason I am not giving examples is that I come in peace to offer access to some of the positions on these topics and have no wish to stir it up just now with the people whose names are on the tip of my pen.

The study of the social relations of science, technology and medicine has become a fairly complex field, and the scholars in the relevant domains would dearly love to have their work read and engaged with by working scientists according to standards which are normal between scientific specialists. Indeed, the debate on the problematic role of science and of various sorts of experts has become immensely important. It is a central theme in current cultural debate, occupying the attention of leading social commentators (see, for example, Daniel Bell's *The Coming of Post-Industrial Society* and Daniel Boorstin's *The Republic of Technology*). Serious engagement between those who think about science has, as a consequence of the significance of the issues, become urgent. If there is to be a responsible and fruitful debate about the appropriate relationship between science, technology and medicine on one hand and wider social, political and economic priorities on the other, surely it behoves all of us to know something about the existing research and points of view. I shall now mention eight approaches and ways into their literatures, without attempting to be exhaustive or to imply that they are mutually exclusive.

DIVERSE APPROACHES

Sociology of Science
This is the most widespread and best-established approach – a recognized subdivision of sociology, differing only in its object of study. The sociologist of science investigates the social forces

at work on and within the general scientific community as well as particular disciplines: growth of fields, funding, institutionalization, professionalization, career structure, patronage, priority disputes, communication networks, elites. The social system and the social relations of science are studied and interpreted according to the currently reigning approach in sociology, which concentrates on the structures and functions which go to make up stability and change. Fieldwork, quantitative methods and studies of the literature are employed, and the special normative features of science, e.g. objectivity and peer review, are given particular attention. The epistemological status of scientific findings and theories is not part of the domain of the sociology of science but is the concern of the philosophy of science (and to some extent of the sociology of knowledge).

The doyen of the sociology of science, who has contributed a stream of important papers since his pioneering work in the 1930s, is Robert K. Merton. His main papers are collected (with an introduction pointing to other approaches) in *The Sociology of Science*. The comparable figure in the sociology of medicine is Eliot Friedson, whose *Profession of Medicine* broke new ground in the study of the professionalism of expertise. *Social Studies of Science* and *Technology and Culture* are periodicals which canvass this approach.

Sociology of Knowledge
The social origins of ideas and the social interests served by them is the domain of the sociology of knowledge, which some consider to be a branch of sociology and others treat as the furthest extension of epistemology into social studies. The recent debate on the problem of making any demarcation between science and ideology arose within the set of issues which developed from the work of Karl Mannheim, whose *Ideology and Utopia: An Introduction to the Sociology of Knowledge*, published in 1936, is the classic treatise. The writ of the discipline ran only to the borders of mathematics and natural science, but it began to be pushed farther in the wake of a broader approach stimulated in 1967 by Peter Berger and

Thomas Luckman's *The Social Construction of Reality*, which emboldened some to ask if conceptions of natural reality might not be constructed in ways analogous to how societies and individuals come to experience the social world. The inviolability of natural knowledge to social analysis has also been challenged by Thomas Kuhn's studies of the social process of conceptual change in *The Structure of Scientific Revolutions*, by David Bloor's work on the sociology of mathematical knowledge; and by the work of his colleagues, Barry Barnes and Steven Shapin, at the Edinburgh Science Studies Unit.

But the most original and detailed study in this tradition is a *tour de force* by Paul Forman. His paper, 'Weimar Culture, Causality and Quantum Theory, 1918–1927: Adaptation by German Physicists and Mathematicians to a Hostile Intellectual Environment', shows just how deeply the sociology of knowledge can penetrate into the cultural evocation of scientific concepts and (in this case) whole approaches to natural phenomena. The issues raised by the sociology of knowledge are always in danger of undermining the foundations of the claims of science to value-free objective knowledge, and there is a large and fraught literature concerned with shoring up those foundations – something that is much easier to do in the physical than in the biological and social sciences.

Anthropology of Knowledge
It has occurred to some researchers to take a different perspective from that of Anglo-American sociology and the related problems of the epistemological status of scientific knowledge. Instead, a small band of anthropologists has suggested that we examine science, technology and medicine in the same way that we study the belief system, social system and process of socialization of any tribe. This approach was inspired by Émile Durkheim and Marcel Mauss's *Primitive Classification* in 1903, and pioneered in Robin Horton's 'African Traditional Thought and Western Science'. It has been most boldly and penetratingly developed by Mary Douglas in a series of studies beginning with *Purity and Danger*. Starting with analyses of concepts of clean and dirty, pollution and taboo, her work (and

that of others influenced by it) has gone on to consider the anthropology of the relations between the symbolic and social orders with respect to mathematics, geology, medicine, environmental debates, and general conceptions in sociology and biology – e.g. the historical and social relations among concepts of health and disease, adaptation and maladaptation, adjustment and deviance. In later work on *Natural Symbols*, *Implicit Meanings* and *Rules and Meanings*, she develops the thesis that we need to look again at the social basis of knowledge, to see how narrow the gap is between the construction of everyday knowledge and of scientific knowledge, and to treat them as a single field of enquiry. She concludes that nothing – no matter how detailed or how abstract and general – escapes the structuring of the social world: all is mediation.

Base and Superstructure
But if all is mediation, what is it a mediation *of*? Are the forces which provide the dynamic of the histories of science, technology and medicine to be found entirely within the internal history of discoveries and the history of ideas? Or is the determination to be found elsewhere in the last instance? Since the nineteenth-century writings of Marx and Engels on ideology and political economy, Marxists have argued that the economic structure of a society is constituted by the contradictory unity of the social relations of production and the material forces of production. This, in turn, provides the base on which arises a legal, political, cultural and intellectual superstructure. Whether or not – and if so how – to include natural science in this scheme has been a hotly controverted question within Marxism and between Marxists and non-Marxists. Science as productive force? Science as superstructure? Beginning with certain exploratory speculations of Engels in *The Dialectics of Nature* and *Anti-Duhring* and extending into the writings of Soviet Marxists, a position has been worked out which treats scientific developments as direct reflections of changes in the economic sphere of production.

The classical expression of this position was made during the dramatic appearance of the Soviet delegation at the 1931

International Congress of the History of Science and Technology in London (proceedings published as *Science at the Cross Roads*). The paper which caused the greatest stir and which remains the *locus classicus* of the 'base-superstructure' (sometimes — I think rightly — called economism or the 'vulgar Marxist') view of science is Boris Hessen's 'The Social and Economic Roots of Newton's *Principia*', which links the main areas in Newton's physics with those in the economy in a one-to-one fashion: 'If we compare this basic series of themes with the physical problems which we found when analysing the technical demands of transport, means of comunication, industry and war, it becomes quite clear that these problems of physics were fundamentally determined by these demands' (p. 166). This approach had a profound influence on the treatment of science and its history by, for example, J. D. Bernal, Joseph Needham and J. G. Crowther. For most non-Marxists, base-superstructure theory is equated with the Marxist approach to science and is therefore a relatively easy target for satirical criticisms. Among the periodicals which have taken Marxist economism seriously (but have not restricted themselves to this approach) are *Science and Society* and *International Journal of Health Services*.

Mediation Theory
Towards the end of his life Engels realized that the dangers of economic reductionism were considerable and argued for paying much more attention to the ways in which the base — now defined more broadly as 'the production and reproduction of real life' — gave rise to, and interacted with, intellectual and cultural forces in complex and indirect ways. The base was determinate in the last instance, to be sure, but there were all the other instances (political, ideological), providing buffers, liberal interpretations, more or less enlightened patronage, institutions, cultural lags, unintended by-products — in short, myriad mediations which modify and can even contradict the ruling ideas of the owners of the means of production. Instead of reading off the likely reflections of economic forces in the superstructure, one had to approach the issues more circum-

spectly and consider matters from the point of view of the totality of relations.

Whole areas of study of these mediations grew up, especially in philosophy, literature and other aspects of culture. Leading theoreticians extended the approach to the understanding of nature, as well as to science and technology in a narrower sense. Thus vulgar Marxism was opposed by a richer view, for example, in Georg Lukács's *History and Class Consciousness*, Karl Korsch's *Marxism and Philosophy*, and Antonio Gramsci's *Prison Notebooks*. They wanted to replace the economism and scientism of vulgar Marxism with a more subtle sense of the interrelations and mutual determinations among philosophical, politico-economic and natural categories. For example, Lukács thought of nature as a societal category, while Gramsci treated the concepts of matter and of objectivity as relative to the history of the mode of production.

In a related series of studies extending into the present, writers associated with the Frankfurt School of Critical Theory concentrated on various aspects of cultural control – what Gramsci called 'hegemony', i.e. the organization of consent in a society by means other than overt physical force without the actual power relations becoming apparent. They placed special emphasis on science and technology, scientific and technological rationality and the role of scientific and technological experts in establishing and maintaining exploitative and repressive socioeconomic structures in the midst of progress and plenty. Their work fully incorporated the investigation of science and technology into cultural studies in ways which have not yet been taken up by academic centres where other aspects of culture are studied. Among the relevant writings of this school are Max Horkheimer and Theodor Adorno's *Dialectic of Enlightenment*, Herbert Marcuse's *One-Dimensional Man*, Jurgen Habermas's 'Technology and Science as "Ideology"' (in his *Towards a Rational Society*), Alfred Schmidt's *The Concept of Nature in Marx*, and – closely allied – Alfred Sohn-Rethel's *Intellectual and Manual Labour: A Critique of Epistemology*.

Labour Process Studies

More recently an attempt has been made to combine certain aspects of mediation theory with the analysis of the production of scientific knowledge which treats it by analogy to other sorts of production – as a labour process consisting of 1. work, 2. raw materials, and 3. means of production. This approach moves away from focusing on what is special about scientific knowledge and considers just how much of the social relations and social processes of the origination, reproduction and dissemination of science, technology and medicine are like those of any other manufactured products on the one hand and other areas of culture on the other. Instead of taking an epistemological approach involving a knowing subject and an object being investigated according to a method, it analyses a social process whereby scientific workers transform raw materials into products by means of instruments and procedures.

The model is based more on craft and industry than on pure knowledge and the ivory tower. It also helps one to think of the sorts of social factors which the sociologist of science studies as integral to the production of knowledge rather than contextual. Grant-getting, customer-contract relations, careers, fashions, power and hierarchy are thereby more easily treated as part of a single domain which includes the research itself. Labour-process studies of science, technology and medicine are being conducted. Some early fruits can be found in the Conference of Socialist Economists' pamphlet on *The Labour Process and Class Strategies*; see also the two-volume collection of labour-process studies (Levidow and Young, eds).

Other Approaches

I want to mention but not discuss two other approaches. The first is *structuralist studies* of the relationship between power and the formal structure of knowledge, a perspective developed in a series of books of astonishing virtuosity by Michel Foucault, translated from the French as *Madness and Civilization*, *The Order of Things: Archaeology of the Human Sciences*, *The Birth of the Clinic*, and *Discipline and Punish*. I

84 Anti-Racist Science Teaching

mention his work for the sake of completeness and because it is brilliant, but it is difficult to imagine how one could build in a systematic way on his very idiosyncratic, individualist insights into various aspects of the archaeology of bodies of knowledge in the biomedical and human sciences.

Second, there is a recent trend to set up courses and programmes, units and symposia, and even posts (in a tight market) on *science* (*or technology or medicine*) *and values* (*or ethics*), especially in North America. My own view, after visiting a number of these enterprises, is that they reflect a deep malaise on the part of educational and research institutions about the need to think critically about ethical, sociopolitical, economic and ideological aspects of science, technology and medicine but that they have not on the whole found their way into the literatures and debates outlined above. That is, they are being set upon a rather *ad hoc* eclectic basis and have not yet made serious contact with *disciplined* studies of these issues, even though it is a welcome sign that some of the relevant questions are beginning to be asked within institutions and are finding a place in the curricula.

SOCIAL CONSTITUTION OF KNOWLEDGE

An analogy occurs to me which might help to overcome some of the scepticism many working scientists have about the social constitution of science. When *On the Origin of Species* first appeared, many scientists who could not see the connection between their findings and Darwin's theory insisted that he could not show them the missing links. He replied that they should not merely look at the tips of the branching trees of speciation but also back along the (necessarily fragmented) stems and branches to the trunk and roots deep in the fossil record. The analogy is that when a working scientist wants to know the social relations of knowledge, s/he needs not only to look at this fact or that ultracentrifuge but also at the concepts, the sorts of classifications, the assumptions, the history and the structural congruences between knowledge and the socio-economic order. This is not likely to be any easier to sort out than any other serious matter. Only if it is treated as a matter for sustained and disciplined enquiry, in which the intellectual standards of special fields other than one's own are treated with the same respect one requires for one's own hard-won special expertise, are we likely to discover how societies constitute their knowledge.

No one is silly enough to suggest that particular findings – the boiling point of water, the spin on an electron, or the light polarization of an organic compound – are determined by socioeconomic forces. But many of the approaches outlined above do invite us to look at various levels of the constitution of knowledge by wider historical forces – not excluding, for example, the concept of electron or the classification of the properties of matter. At the deepest level, world-views or philosophies of society are arguably historically constituted. Within a given mode of production different epochs call up different disciplines and topics, along with criteria for acceptable answers to the questions we put to nature. Within a given period different priorities and conceptual frameworks arise.

The undoubted contributions of the internal history of ideas and discoveries can thus be complemented by asking a series of questions about, for example, why the sixteenth and seventeenth centuries (in certain countries) gave us closely intertwined and far-reaching changes in the rise of capitalism, the Protestant ethic and the development of the metaphysical assumptions of modern science; why the late seventeenth century gave us Newtonian mechanics, while Weimar Germany gave us acausality and quantum mechanics; why the eighteenth century was an era of classifications across a wide range of disciplines, while the mid-nineteenth was preoccupied with progress through competition, and the 1950s and 1960s concentrated on micromechanisms of information coding and reproduction; why the early nineteenth century gave us phrenology, the turn of the present century eugenics and IQ, the 1920s behaviourism, followed by the development of ethology and most recently, by sociobiology.

Anti-Racist Science Teaching

References

(All books are published in London unless otherwise noted.)

Barry Barnes, ed., *Science in Context: Readings in the Sociology of Science*, Milton Keynes, Open University Press, 1982.
Daniel Bell, *The Coming of Post-Industrial Society: A Venture in Social Forecasting*, New York, Basic, 1976.
Peter Berger and Thomas Luckman, *The Social Construction of Reality: Treatise in the Sociology of Knowledge*, Penguin, 1979.
David Bloor, *Knowledge and Social Imagery*, Routledge & Kegan Paul, 1976.
Daniel J. Boorstin, *The Republic of Technology*, New York, Harper & Row, 1978.
N. I. Bukharin *et al.*, *Science at the Cross Roads* (1931), Cass, 1971.
Conference of Socialist Economists, *The Labour Process and Class Strategies*, Stage 1, 1976.
Mary Douglas, *Purity and Danger: An Analysis of Concepts of Pollution and Taboo*, Routledge, 1966.
Mary Douglas, *Natural Symbols: Explorations in Cosmology*, New York, Random, 1972.
Mary Douglas, *Rules and Meanings*, Penguin, 1973.
Mary Douglas, *Implicit Meanings: Essays in Anthropology*, Routledge, 1978.
Emile Durkheim and Marcel Mauss, *Primitive Classification* (1903), Routledge, 1963.
Friedrich Engels, *Anti-Duhring*, Lawrence & Wishart, 1955.
Friedrich Engels, *Dialectics of Nature*, Lawrence & Wishart, 1977.
Paul Forman, 'Weimar Culture, Causality and Quantum Theory, 1918–1927: Adaptation by Germany Physicists to a Hostile Intellectual Environment', *Historical Studies in the Physical Sciences*, 3, 1 (1971), 1–115.
Michel Foucault, *Madness and Civilization*, Tavistock, 1971.
Michel Foucault, *The Order of Things: Archaeology of the Human Sciences*, Tavistock, 1974.
Michel Foucault, *The Birth of the Clinic*, Tavistock, 1976.
Michel Foucault, *Discipline and Punish: Birth of the Prison*, Penguin, 1979.
Eliot Friedson, *Profession of Medicine*, New York, Dodd, Mead, 1970.
Antonio Gramsci, *Prison Notebooks*, Lawrence & Wishart, 1973.

Jurgen Habermas, *Towards a Rational Society: Student Protest, Science and Politics*, Heinemann Educational, 1971.
Max Horkheimer and Theodor Adorno, *Dialectic of Enlightenment*, Verso, 1979.
Robin Horton, 'African Traditional Thought and Western Science', *Africa*, 37 (1967), 50–155.
Karl Korsch, *Marxism and Philosophy*, New Left Books, 1972.
Thomas Kuhn, *The Structure of Scientific Revolutions*, Chicago, Univ. of Chicago Press, 1970.
Les Levidow and Bob Young, eds, *Science, Technology and the Labour Process: Marxist Studies*, vol. 1, Free Association Books, 1984; vol. 2, 1985.
Georg Lukács, *History and Class Consciousness: Studies in Marxist Dialectics*, Merlin, 1971.
Karl Mannheim, *Ideology and Utopia: An Introduction to the Sociology of Knowledge*, Routledge, 1954.
Herbert Marcuse, *One-Dimensional Man*, Routledge, 1964.
Robert K. Merton, *The Sociology of Science: Theoretical and Empirical Investigations*, Chicago, Univ. of Chicago Press, 1979.
Alfred Schmidt, *The Concept of Nature in Marx*, New Left Books, 1971.
Alfred Sohn-Rethel, *Intellectual and Manual Labour: A Critique of Epistemology*, Macmillan, 1978.

Anti-Racist Curriculum Change

INTRODUCTION

It is one thing to recognize that Western science has been shaped by racist priorities, and quite another to teach science in a way that challenges racism. Anti-racist teachers face numerous obstacles, not the least of which are widespread notions about what is or is not science, and the myths of political neutrality and objectivity that constrain discussion on science education. This section presents five case studies of how science has been or could be taught in ways which acknowledge both its ideological content and political context.

Liz Lindsay's critique of traditional classroom approaches to nutrition and hunger demonstrates that they are founded on and likely to perpetuate common ignorance about the causes of Third World poverty, and are often based on racist prejudices. She counterposes her own approach, which explores issues of power and control over what crops are grown and how they are distributed.

Michael Vance examines the racist underpinnings of the biology curriculum, and argues that teachers should use classroom discussions to deal openly with students' preconceptions about so-called 'racial differences'. He shows that even supposedly anti-racist biology textbooks fail to mount an adequate challenge to racist stereotypes. He poses a question for discussion: Why does biology teaching in this society emphasize some differences among individuals rather than others?

The next article takes the biology example further by showing how multicultural approaches focus on 'racial' or 'cultural' differences at the expense of challenging institutionalized racism. The authors point out that multicultural education is not an antidote to racism, and that to present it as such is diversionary: multicultural education frequently provides a cloak for the racism which it would claim to oppose.

For many schoolchildren, images of East African game reserves symbolize the supposed need to protect wildlife from (African) people. Malcolm Green shows how Kenya's 'conservation' policies serve Western tourism while alienating the indigenous local people from the land. He demonstrates the racist nature of those policies and challenges their imposition of Western values. He suggests that conservationists could learn from the African people's long experience in wisely using the wildlife. This article is intended as a starting point for anti-racist approaches to ecology.

The last article in this section deals with the overall attempt at curriculum innovation to provide real equality of opportunity at one school. Realizing that the traditional curriculum structure posed an obstacle to such equality, the staff at Holland Park School decided to integrate their subject departments and try out new teaching methods to place science education in its social context. Here they assess the successes and failures of their attempt, and draw lessons for the future.

92   Anti-Racist Science Teaching

Anti-Racist Curriculum Change

The diseased and the healthy: this textbook portrays black people as examples of sickness and malnutrition; it emphasizes the consequences of poverty, while ignoring its causes. The same book portrays white people as examples of healthy achievers.

NUTRITION AND HUNGER: TWO CLASSROOM APPROACHES

LIZ LINDSAY

Most biology textbooks deal with Third World poverty through references to malnutrition, disease and starvation. Although these problems manifest themselves biologically, their sources lie in issues of geography, history, politics, economics and sociology. An integrated approach to the study of malnutrition and deficiency diseases is essential if they are to be understood. The following case study of classroom practice contrasts the more traditional approach to nutrition with an anti-racist approach; it argues that the latter is preferable in terms of pedagogy and intellectual honesty.

THE TRADITIONAL APPROACH

Michael Vance, of West Indian origin, describes his recollection of his experiences of secondary school science in the late 1960s and early 1970s as 'one of receiving images of a Third World rife with disease and needing the science of the West to combat its problems'.

In most texts, Third World countries are used to demonstrate the effects of starvation, gross malnutrition and disease no longer suffered by the more affluent nations. Most texts accept without question the superiority of everything European/American, including farming methods, food processing and distribution, and the Western diet. Third World populations are blamed for their own problems.

Many teachers accept the charity fund-raising type of material, which presents the Third World as societies in chaos and poverty, desperately requiring help from the 'saviours' of the West. Supposedly these Western elites alone can solve Third World problems with their superior understanding of the situation and provision of Western aid, such as experts, seeds and tractors.

Students do not learn that prosperous countries, comprising 25 per cent of the world's population, eat two-thirds of the world's food production or that much of the food imported by the affluent countries is produced by the poorer nations. Nor do the students learn about the large land areas now being used in the Third World to produce these cash crops for export, or

Increasing food production

Textbooks present Western technology as the solution to food shortages. But in the Third World such technology – and the accompanying fertilizers, pesticides and 'miracle' seeds – involves a shift towards cash crops for export.

about the exchange rates which increasingly tend to favour the industrialized nations. Those rates mean that Third World countries receive far less than before for their goods, even if they produce more.

Many school textbooks offer only one explanation for the suffering of the 'starving millions' in the Third World – their overpopulation and their failure to use adequate birth control methods. The assumption is that these problems can be overcome only when Western scientists devise cheaper birth control methods which people in the disadvantaged countries can be persuaded or forced to use effectively.

Some texts also suggest a technological solution for malnutrition in the poorer nations – Western manufacture of a cheap form of protein, using the West's waste materials or by manufacture of single-cell protein (George, 1976, p. 273). Such texts ignore the problem of distribution; the malnourished will continue to starve because they are landless, jobless and unable to purchase even the waste food of the West. Furthermore, the texts ignore problems of food production in a cash crop economy, in which land is used to grow crops for export, instead of for the local people.

AN ANTI-RACIST APPROACH TO BRITAIN

In the anti-racist approach that I am developing, students work in groups, using a variety of sources. They explore various aspects of the topic, discuss their findings and make a critical evaluation of each other's consclusions. With this resource-based learning, it is possible for members of the class to work independently and thus release the teacher to have discussions with individuals or groups in a non-threatening atmosphere.

After a general introduction to the ecosystem, students research different types of farming in Britain in order to compare the effects on the ecosystem of the modern monocultural farming with more traditional mixed farming. They pay special attention to the latest high-energy and capital-intensive American system, which is usually regarded as a model by the rest of the world. They investigate the disadvantages as well as the merits of the Green Revolution, still hailed in many school texts and films as the great technological solution for world hunger.

They test foods such as potatoes, cereals and pulses for protein content. They consider figures which give protein concentration, quality and yield/acre for these staple foods, comparing these figures with those for animal protein. They consider the amount of energy required to produce the same amount of food value in animal and vegetable protein. As a result, they can appreciate the cheapness and energy efficiency of a vegetarian diet.

They also look at the reasons for and the effects of food processing. They connect that to the growing popularity of the cheaper restaurants or 'takeaways', with their quick service of foods with high salt, fat and sugar content. There is also the more common use of packaged foods, often artificially coloured or flavoured, attractively packed and frequently advertised. They consider what food values as well as fibre are destroyed in food processing, which so greatly increases the price.

Students visit a supermarket to find out the country of origin of the various foods, thus obtaining first-hand evidence of the variety provided in Britain's diet by Third World countries. By costing for the minimum daily intake for a good balanced diet as defined by their textbooks, students can see why some low-income groups in Britain suffer from malnutrition and poor health. Income figures are provided for low-paid workers, pensioners and the unemployed, as well as for better-paid workers.

Students are given data for the difference in diet of the different socioeconomic groups, as well as for the incidence of heart, circulatory and digestive disorders in each group. They find that in Britain the poorer socioeconomic groups suffer most from these ailments. Moreover, these groups buy smaller quantities of fresh fruit and vegetables, which are more expensive and less filling than cheaper foods with little fibre and high proportions of fat, sugar and starch.

An excellent example of ways in which government intervention can improve general nutrition and morale was shown during World War 2, when the maintenance of a healthy population was considered vital to the war effort. Scarce foods and tea were rationed and subsidized to ensure that every member of the population had access to a share of basic foods. At the same time a government regulation, requiring manufacturers to add vitamin D to all margarine, virtually eradicated rickets, the long-standing 'English disease', especially prevalent among the poorer population. A higher incidence of rickets is evident today among some of the non-margarine-eaters in the Asian population in Britain, but in this case no government action has been taken; such people can eliminate rickets only if they supplement their diets, at their own expense, with vitamin D pills. This may be interpreted as a case of institutional racism, since the diet of white British people is supplemented, at public expense, but not that of black British people.

MALNUTRITION IN THE THIRD WORLD

The subject of malnutrition, disease and starvation in the Third World cannot be fully understood without an integrated approach including geography, history, politics, economics and sociology. The following example, however modest, is a great improvement on the meagre and misleading treatments in biology textbooks.

Before I begin a study of the incidence and effects of malnourishment and famine in the Third World, I ask students to write down their own ideas of the causes, which they keep for reference. Not unexpectedly, with varied backgrounds in terms of ability, socioeconomic class and national origin, they express a wide range of views about the causes. These often include negative stereotypes, with their racist overtones of laziness, shiftlessness, stupidity, ignorance, lack of skill and overbreeding. Sometimes students also mention poor soils and oppressive or unpredictable climates. These tend to be the assumptions of students taught by the more traditional method.

While some students (or their relatives) come from Third World countries, a greater number have links with Ireland, so I use the Irish potato famine of 1846–50 as a starting point. Students have no difficulty in discovering that two million people, 25 per cent of the Irish population, died from starvation during these years, while another 25 per cent migrated. The only crop to be affected by blight was the potato crop. Farmers continued to produce cereals, cattle, pigs, eggs and butter – enough to feed twice the Irish population over the period – all of which were exported. Farmers were usually forced to sell all their crops other than potatoes to pay their rent. Failure to do so meant eviction and the razing of their cottages; they could find nowhere else to live or work.

Since the successful exported crops were grown by the Irish farmers, the Irish famine cannot be attributed to the laziness, shiftlessness, ignorance or lack of skill on the part of the farmers, to the poorness of the Irish soil or to the vagaries of the climate. Nor had the population increased suddenly before or during the famine.

Having studied the Irish famine, all students, especially those with Irish roots, are better able to understand and empathize with the hungry in the Third World. Later they can see resemblances between Ireland and the Third World, as examples in the history of colonialism. Working in groups, students then try to test their original hypotheses regarding the reasons for world hunger. They use a wide range of books and pamphlets with conflicting interpretations of the causes of and solutions to world hunger. This encourages critical evaluation and lively discussion. The extent of teacher guidance given depends on the maturity and the ability of the students.

Students read the observations of visitors to Third World countries before they felt the full impact of Western colonization, trade or 'aid'. Many parts of Africa, Asia and other Third World countries were described as ordered societies. Their traditional farming methods, with careful conservation, had generally produced an adequate supply of food over thousands of years. The poor suffered most during periodic famines, but those did not have the scope of present-day famines.

By a variety of means the European colonizers were able to induce the populations of the Third World to sell or use their most fertile land to produce cash crops for export instead of producing food for themselves. Students have already noted in supermarkets a tremendous quantity and variety of imports from Third World countries – tea, coffee, sugar, cocoa, peanuts, rubber, cotton, sisal, jute, tropical fruits and even flowers. For example,

> Worldwide, 250,000 sq miles – one and a half times the area of California, enough land to feed the world's present hungry people several times over – is given to cash crops – like coffee, tea, cocoa and the rest, of which many have little or no nutritional value and much of what does have nutrition value . . . is fed to livestock, which squanders about 90 per cent of the original vegetable protein. (Tudge, 1977, p. 14)

> Today cash crops occupy 55 per cent of the best land in the Philippines, 80 per cent in Mauritius, and peanuts alone occupy 50 per cent of the land in Senegal. (George, 1976, p. 39)

In most Third World countries an increasing amount of the best arable land is being used for cash crops. Production of food for local people is being confined to smaller areas of land, which are intensively cultivated and overgrazed. The result is soil exhaustion, erosion and greatly decreased productivity, as depicted in the video 'Haunt of Man'. Thousands who once farmed land to produce their own food are now landless, homeless and jobless. Jobs in secondary and tertiary industry are very limited, since the wealthy countries severely restrict access to their markets of goods manufactured in the Third World.

Concern about this change in land use was expressed in a letter to the French government by a French colonial inspector who wrote from famine-stricken Upper Volta on 7 March 1932:

> One can only wonder how it happens that populations who always had on hand three harvests in reserve, and to whom it was unacceptable to eat grain that had spent less than three years in the granary, have suddenly become improvident. They managed to get through the terrible drought years 1912–1914 without hardship ... now these people, once accustomed to food abundance, are living from hand to mouth. I feel morally bound to point out that the policy giving priority to industrial cash crops has coincided with an increase in the frequency of food scarcity. (George and Paige, 1982, p. 42)

While some smaller farmers grew the new cash crops, colonists acquired land by various means and established large plantations that they managed themselves, their workers coming from communities who had lost their land. While larger holdings may be more capital-intensive with greater productivity per person-hour, the smaller, more traditional farms are more labour-intensive and usually have greater productivity per acre.

Ecological damage and diminishing returns have resulted from the continued use of the same land for the same crop and from intensive use of chemical fertilizers and pesticides. Unlike the USA, the Third World cannot afford to squander its land. Some Third World countries find that they can no longer produce enough food for their own needs and are having to import food, even though the cost of this food may exceed the amount they receive for cash crops. Unfortunately, the food imported tends to be of the packaged, processed type with additives such as sugar, very much more expensive and far less nutritious than the traditionally grown food. The wasteful and expensive American diet does not have to be accepted as the model which others should emulate. Some communities which are still growing and eating their traditional food prevent protein deficiencies by mixing their vegetable proteins. In the Andes they mix cornflour and bean flour; the Chinese, for centuries, have added Chinese cabbage and soy beans to rice.

Third World countries have no pensions for the old, sick or unemployed. Those without money cannot afford to buy food. Students can try to imagine the situation in Britain if all pensions were withdrawn. As another comparison with Britain, students have no difficulty in seeing that those who are malnourished are increasingly susceptible to illness and disease. Nutritional deficiency diseases are still seen throughout the world, although starvation is now limited mainly to the Third World.

This juxtaposition in a school textbook conveys a message: 'overpopulation' causes poverty, not the other way around.

Birth control is not willingly practised in the poorest countries. The main reasons are the high death rate among children and the need for some security in sickness or old age. Students can find statistics to show that whenever the birth rate has decreased in Third World countries the average income has first increased, as has occurred in Britain.

Textbooks and the media often express fears that world food resources will be inadequate because of the continued population increase in the Third World. The industrialized world need have no fear that they or their livestock will suffer; throughout history, in times of shortages the most powerful survive, while the poorest perish.

And what happens to the food now produced? The US at present, with 6 per cent of the world's population, consumes 35 per cent of the world's resources. Rich countries as a whole, 25 per cent of the world population, eat two-thirds of the world's food production. 'The average US American gets through about 1,800 lb of grain a year but eats only 200 lb as grain, the rest being passaged through livestock, while the average Third Worlder gets about 500 lb a year, mostly eaten in pristine form' (Tudge, 1979, p. 19). A change in popular preference to lean meat rather than grain-fattened marbled beef could improve the health of the affluent countries and quickly provide a grain surplus for consumption by the Third World countries.

However, there is no world shortage of food at present. Since 1960, world grain production has been growing faster than population. Improvements in technology have led to a 60 per cent increase in average yields. As one journalist described the problem in terms of distribution:

> The fact that the world has entered an era of permanent grain surpluses is perhaps hard to appreciate when TV screens are filled with pictures of starving African children. To put that human tragedy into perspective, it has to be understood that the entire shortfall in Sahelian grain this year – the harvest which would have eliminated the famine – has been estimated at 3–4 million tonnes. World grain

surplus stocks currently stand at 190 million tonnes. (Walsh, 1985)

Many affluent countries are restricting production because they cannot market their food surpluses. Similarly, during the Great World Depression of the 1930s the unemployed in the industrialized countries were severely undernourished, ill-clothed and ill-housed, but not because the farmers could not produce enough food, or the factories enough goods. Indeed, the American farmers were paid by the government to destroy crops to keep prices up. The unemployed suffered simply because they could not pay for food, clothing and housing, especially given the absence of even subsistence-level unemployment benefits.

Today, even an increase in crop yields in the Third World is not in itself the solution. While the volume of crops has increased considerably, the prices paid to the underdeveloped countries by the affluent countries for the cash crops continue to drop compared with the prices that have to be paid for imports. In other words, countries have to produce more and more cash crops to get the same goods in return. Examples:

> 5 tons of sisal bought a tractor in 1963;
> 10 tons of sisal were needed in 1970;
> 12.5 tons of rubber bought one tractor in 1963;
> 37.5 tons of rubber were needed in 1970.

The 'actual volume of exports by the developing world has increased by over 30 per cent in the last twenty years whilst their value, in real terms, has increased by only 4 per cent' (George, 1976, p. 37). There could be no solution unless the more affluent nations increase the volume of the goods they export in return for the products they import from the Third World.

As was stated in a United Nations debate, before all the starving in the Third World have the slimmest chance of eating enough, 'it would be necessary to break the vicious circle of unemployment, low food production – low productivity – low income which strangles such a large part of mankind. This, in

Nutrition and Hunger 105

> ONCE UPON A TIME, BEFORE OUR LAND WAS USED FOR COFFEE, SUGAR AND TEA, WE USED TO GROW FOOD FOR THE PEOPLE.

turn, might entail very deep transformations in present socioeconomic structures' (George, 1976, p. 43).

I find that students become very engrossed in this topic, read widely and become very involved in discussion. They frequently express surprise at the alternative viewpoints and conflicting interpretations, I think that the experience helps students to become more critical in reading about science and less willing to accept explanations separated from the broader social and political context. They tend to see that alleviation of hunger in the Third World does not depend on the production of more food or newer types of food in the industrialized countries, nor even an increase in crop yields in Third World countries, unless the poor, landless and unemployed can afford to pay for it.

Any solution requires an international and national redistribution of power, land, employment and resources.

Suggested Reading

(All books published in London unless otherwise noted.)

Charlie Clutterbuck and Tim Lang, *More Than We Can Chew: The Crazy World of Food and Farming*, Pluto, 1982.
Stephen Croall and William Rankin, *Ecology for Beginners*, Writers & Readers, 1981.
Barbara Dinham and Colin Hines, *Agribusiness in Africa*, Earth Resources Research, 1983.
Lesley Doyal, *The Political Economy of Health*, Pluto, 1979.
'Eat Your Heart Out', video available from Concord Films, London.
Susan George, *How the Other Half Dies: The Real Reasons for World Hunger*, Penguin, 1976.
Susan George and Nigel Paige, *Food for Beginners*, Writers & Readers, 1982.
'Haunt of Man', BBC Horizon film (available from ILEA film library).
Teresa Hayter, *The Creation of World Poverty*, Pluto, 1981.
Barbara Insel, 'A World Awash with Grain', *Foreign Affairs Quarterly* Spring 1985.
New Internationalist magazine.
Colin Tudge, *The Famine Business*, Faber, 1977; Penguin, 1979.
Michael Vance, 'Biology', in A. Craft and G. Bardell, eds, *Curriculum Opportunities in a Multi-Cultural Society*, Harper & Row, 1984.
M. Walsh, *Sydney Morning Herald*, 2 May 1985.
C. Wardle, *Changing Food Habits in the UK*, Earth Resources Research, 1977.

BIOLOGY TEACHING IN A RACIST SOCIETY

MICHAEL VANCE

Biology, together with other science subjects, has so far remained relatively untouched by the changes which will be necessary if schools are to tackle racism in the curriculum. In schools with high proportions of black pupils, some science teachers question the traditional approach to science teaching. At present there are more questions than answers. There is a great need for staff discussion; this may lead to initiatives which should permeate the curriculum. There is need for a biology curriculum more appropriate to the needs of the pupils, the more politically aware of whom may well feel that learning the classical biology curriculum is tantamount to colluding with racism. My experiences show that this feeling is widespread among black pupils in North London schools.

All children in Britain are growing up in a multicultural but racist society. Ideally, their school experience should lead them to appreciate the multicultural and racist origins and manifestations of knowledge; it must lead them to see beyond their own culture. It must prepare them to respect, value and defend cultures which are continually under racist attack. Education in biology has a major role to play in establishing a framework where enquiry can challenge racist images and practices – not least because these may be based on ideas which appear to have scientific justification, and permeate biology texts.

Pupils' questions about different human attributes and physical form should be dealt with in biology, not only when they happen to arise, but as part of a coherent anti-racist policy throughout the school curriculum. Otherwise stereotypes will continue to be reproduced by an unchallenged racism. Such a policy must not be confined to institutions with black pupils, but needs to include schools where pupils are predominantly white; racism is a problem in these schools, too, since all schools in Britain operate within a racist society. Schools in the white suburbs and rural areas must not remain exempt from the initiatives taken up in the inner-city schools.

CLASSROOM EXPERIENCES

Biology teachers have to be on their guard against images received by pupils from texts which portray black people as simply primitive or undeveloped, or which present 'development' as synonymous with becoming more like the West.

As someone of Caribbean origin, educated in England during the late 1960s and early 1970s, my experience was one of receiving images of a backward and deprived Third World rife with disease and needing Western science to combat its problems. Today, in the 1980s, the biology curriculum has yet to redress this imbalance. By providing negative images of the poverty-stricken 'Third World', which cannot cope with rickets, kwashiorkor or beri-beri, the curriculum prepares pupils to take for granted the dominant British culture. It does not provide a framework whereby they can make decisions about their own culture – whether this be the dominant culture or a so-called minority culture.

As a school student during the awakening of a political black consciousness, I directed my main focus on the ways whereby black people could have power and control over their own lives. The individual or collective role within this cannot be found unless we understand the cultural domination of one group by another (Carmichael and Hamilton, 1967). In a racist society, black students may wish to find some explanations of their origins and physiology within the biology curriculum: they may wish to spend some time in history and economics lessons learning about the reasons for black presence in Britain, and about the historical and political roots of racism. Where possible, differing views within cultures should be expressed,

Biology Teaching in a Racist Society

including the pupils' own. This is not without its difficulties, nor is it necessarily anti-racist. There are difficult areas where Western culture clashes with others in the science classroom, and where pupils expect teachers to take sides.

Once, while discussing circumcision, the classroom debate focused on the practice of female circumcision – itself a euphemism which is often used to mean infibulation (removal of the external and internal labia of a woman's genitals), as well as removal of the clitoris. This practice aroused strong feelings in the pupils, who felt that this 'circumcision' was wrong; the class were united in their outrage. I felt it necessary to extend the discussion by showing them a film on female circumcision, which included discussion with a Nigerian woman who supported the practice. The black girls of the fifth-year class saw that many African women have a very different view of their own bodies and that we should beware of culturally chauvinistic judgements about other people.

Female circumcision is an issue about which there is dissent within the African cultures. Many women – and men – regard 'circumcision' as an extreme form of male domination which deprives women of the enjoyment of their own bodies, for the sake of supposedly giving more pleasure to men. Other people disagree and regard the practice as a necessary and important part of their culture. Anti-racist educators need to be careful in treading the line between respecting people's cultures and adopting an uncritical attitude towards culturally linked forms of oppression. It is necessary to educate about dissent *within* cultures, as well as between them (MacLean, 1980).

On another occasion, an illustration from a commonly used text gave rise to a similar situation in which it became necessary to challenge the racist thinking of students. The diagram showed a man passing eggs of the blood fluke *Schistosoma mansoni* into a river by urinating, while other men were swimming in and drinking the river water. Donna, a girl of Caribbean parentage, said, 'Those Africans are nasty, pissing in the water.' We then had a short discussion about the origins of our own drinking water. Modern industrial contaminants and sewage pollute some of Britain's rivers. The water in

Transmission of *Schistosoma mansoni*

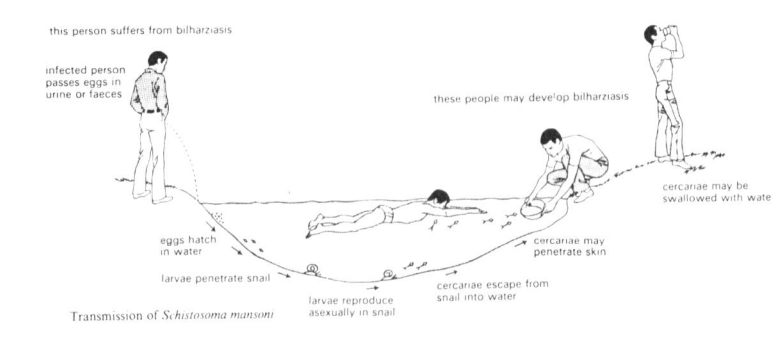

- this person suffers from bilharziasis
- infected person passes eggs in urine or faeces
- these people may develop bilharziasis
- cercariae may be swallowed with water
- eggs hatch in water
- cercariae may penetrate skin
- larvae penetrate snail
- cercariae escape from snail into water
- larvae reproduce asexually in snail

British rivers is just as polluted as the river water from other parts of the world. It was interesting to discuss how the implications of the diagram confirmed Donna's own view of Africa and Africans, despite the fact that the diagram was not labelled and could have referred to anywhere in the world. This incident said as much about the negative images of Africa presented in the media as it did about anything else: 'If it's nasty, it must be African, and if it's African, it must be nasty' was Donna's assumption.

How is it that black British children learn contempt for anything to do with Africa? It is interesting to note here that when black British children wish to insult each other, 'African' is a term which does the job better than any other. When we are teaching, we need to be aware that we are not working in a neutral situation. Pupils come to biology with preconceptions about the world and about black people. It is these which help to condition their perceptions of what they see, read and discuss in a biology lesson. If relatively innocuous pictures can reinforce racist attitudes, the ones which are less so may be much more dangerous. The images in biology textbooks may be far more potent in promoting racist interpretations of poverty than the relatively innocent river fluke picture was in confirming Donna's views of Africa.

In the case described above, the incident stimulated the pupils to re-examine Britain and Africa and eventually to reconsider their own images of Africa. They discovered that, where water is untreated, people in the Third World countries normally boil it before drinking. To supplement the discussion on this issue, we 'plated' samples of tap water and grew bacterial colonies on the agar; the class realized that even in 'advanced' Britain water does contain micro-organisms. This helped the students to question their assumptions. Such practical work can be used to use the mixed experiences and examine the prejudices of any class. This means introducing pupils to different ways of life, awakening an informed interest in themselves and other people, while at the same time challenging racist attitudes and their origins.

For this approach to succeed, it is essential that the teacher be aware of racism and determined to tackle it wherever and however it is expressed. One strategy for combating the dominant ideology, especially with inadequate teaching resources, is to use pupils' own experiences and to involve parents, community agencies, etc. Such an approach may help the pupils from 'minority' backgrounds to overcome the classroom alienation that they would otherwise experience. Thus they can feel that they have something to contribute; this allows them to begin to feel their own strength, thereby increasing motivation and participation. It may have been possible, for example, to get an African parent to talk to the pupils about the standards and methods of hygiene, and to explain why it is that some countries are unable to afford expensive water purification systems. It is likely that such a person would mention Africa's great wealth, much of which finds its way to other parts of the world while most African people remain poor. To examine an issue such as schistosoma within a political, economic and historical framework would have been useful: to get a parent's help in this would have been more so, in that the school could have been seen to value the contributions of ordinary people within the community.

Some teachers feel that using black pupils or their parents in this way is both tokenistic and voyeuristic: they may become mere visual aids. However, every situation is unique, dependent upon the personal history of the kids, the strength of feeling and organization of their communities. For example, a teacher in my school commented that in her form the Asian pupils felt very uncomfortable when asked to discuss preparing food at home and refused to contribute; in fact the rest of the class were united in their distaste for curry as 'Paki' food. Another teacher taking a parallel class for science found that three Asian pupils in his class came forward easily to talk about the grinding of spices, etc. White pupils in the class commented that they loved curry and used to get takeaways from a local shop. The teacher felt that the Asian pupils' confidence came from a secure atmosphere nurtured by outspoken members of the class supporting them in a number of ways, including wearing GLC 'I'm against Racism' badges.

The level of overtly expressed racism can vary from class to class. Participation of black and ethnic minority pupils can sometimes be encouraged if the school identifies 'racism' as a white problem and looks for the solution amongst white pupils rather than always using black children as the focus of attention. Of course, the idea that racism is a white problem is one that itself needs to be examined critically. White people were not born racist; they were taught to be so – just as many black people have also been taught to be so. In a racist society, racist attitudes and behaviour are learned; they are internalized by black and white people alike. It is necessary to examine how – and why – if the curriculum is to be a means by which racist ideology and practice are challenged, instead of a means by which they are perpetuated.

GENETICS AND 'RACE'

Pupils have ideas about each other which they derive from their parents and their culture. These often include racist views, some of them apparently based on scientific ideas. It is not uncommon to find pupils and teachers who believe that white people are more intelligent than black people, or that Blacks are better at running than Whites. Other stereotypes include:
☐ Black children have better motor skills than white children.
☐ Black men have bigger penises.

Biology Teaching in a Racist Society

☐ Black people have stronger body odours.
☐ Black people are more aggressive.
☐ Black people have greasy skins.
☐ Black people are more closely related to apes, and they look more alike than Whites.

Pupils with such ideas, which they believe to be scientific, rarely find them challenged by contemporary biology courses. If pupils' racist ideas are not tackled, then the school has failed; it has colluded with racism. *Bulldog*, the youth newspaper of the National Front, provides a reading list for young fascists. It includes the work of Lorenz, Tinbergen, Eysenck, Skinner and Morris. All these authors are also on the reading list for the Nuffield Advanced Biology course. Incidentally, some of these names will also be familiar to teachers who are not biologists, as they are required reading on some of the 'Certificate in Education' courses by which teachers qualify for their profession (e.g. PGCE).

That the work of these writers lends itself to fascist use is clearly illustrated in the following quotation from Konrad Lorenz, who wrote during the Nazi era:

> This high valuation of our species – specific and innate social behaviour patterns – is of the greatest biological importance. In it, as in nothing else, lies directly the backbone of all racial health and power. Nothing is so important for the health of a whole *Volk* as the elimination of 'invirent types': those which, in the most dangerous, virulent increase, like the cells of a malignant tumour, threaten to penetrate the body of a *Volk*. This justified high valuation, one of our most important hereditary treasures, must, however, not hinder us from recognizing and admitting its direct relation with nature. It must above all not hinder us from descending to investigate our fellow-creatures, which are easier and simpler to understand, in order to discover facts which strengthen the basis for the care of our holiest racial, *volk*ish, and human hereditary values. (quoted in Kalikow, 1978, pp. 174–5)

Anti-Racist Science Teaching

Lorenz feared that in conditions of civilization where natural selection was inoperative, 'the increase in the number of existing mutants' might lead to an 'imbalance of the race'. If so, 'then race-care must consider an even more stringent elimination of the ethically less valuable than is done today, because it would, in this case, literally have to replace all selection factors that operate in the natural environment' (Kalikow, 1978, p. 176).

Lorenz has been a best-selling author and a Nobel Prize winner; that gives some indication of the scientific 'respectability' and popularity of his views. Fascists consider the work of Lorenz, Tinbergen (another Nobel Prize winner), Eysenck, Skinner and Morris as a necessary part of fascist ideology. Those writers are cited to legitimize such ideology as scientific. To my knowledge, none of those writers has ever repudiated such a use of their work.

The National Front made three major interventions outside the school gates during the 1977–8 school year and again in 1983 before the General Election; in 1984, 1985 and 1986 their activities intensified. These interventions have included leafleting and distributing *Bulldog*. The first two issues of this race-hate paper contained articles on 'race mixing', 'the Jews' and 'the race problem'. Their ideology is perhaps summarized by the following passage:

> The claim that integration is harmless and that all races are equal is totally false. Races differ, not only in the colour of their skins, but in other physical ways and especially in temperament and innate intelligence. Scientists tell us that these differences are not the result of environment only, but mainly of heredity. 'Men are not born equal,' says Professor Francis Crick, a Nobel Prize winner ... It is not bigotry to oppose multiracialism. It is a natural healthy instinct to preserve one's own kind. A leading scientist, Sir Arthur Keith, has said that what is called prejudice is simply nature's way of improving mankind through racial differentiation.

Thus the passage promotes racism as a scientifically proven theory and therefore as 'truth'.

Children tend to receive uncritically the racist views which their parents themselves may have internalized as children. Within the current crisis of capitalism – with its rising unemployment, deteriorating housing, etc. – the political conditions for the rise of racism and fascism are enhanced. This makes the need for anti-racist education greater. So-called 'racial differences' cannot be treated as unproblematic.

'RACE' IN TEXTBOOKS

Nuffield Biology Text 4, *The Perpetuation of Life*, is part of a well-established course for thirteen- to sixteen-year-olds. It gives great attention to questions of race, heredity and intelligence. It commences by raising questions about the myth of social superiority. Quoting the famous 1951 UNESCO statement on equality, it goes on to reinforce this statement by claiming:

> Some biological differences between human beings within a single race may be as great as or greater than the biological differences between two races ... Available scientific knowledge provides no basis for believing that the groups of mankind differ in their innate capacity for intellectual and emotional development ... Social changes have occurred that have not been connected in any way with changes in racial type. Historical and sociological studies thus support the view that genetic (that is, inherited) differences are of little significance in determining the social and cultural differences between groups of men. (p. 19)

Unfortunately, the UNESCO statement attempts to combat racism by using language which legitimates the notion of 'racial type' and masculine 'mankind'. The racist and sexist language of the text is worse than unfortunate, as science books are likely to be held in very high esteem. Unquestioned ideas may be all the more acceptable if presented as part of a science lesson – and one which may seem to be attacking racism. The attack on racism is further tempered by the following:

> Not all biologists were completely happy about this statement. One eminent geneticist had this to say: 'I feel at times that it was bending over backwards to deny the existence of race in the sense that this term has been used for political purposes in the recent past. I of course entirely agree in condemning the Nazi race theory, but I do not think that the case against it is strengthened by playing down the possibility of statistical differences in, for example, the mental capacities of different groups.' (p. 19)

The Nuffield text then quotes another geneticist:

> By trying to prove that races do not differ ... we do no service to mankind. We conceal the greatest problem which confronts mankind ... namely, how to use diverse gifts, talents, capacities of each race for the benefit of all races. For, if we were all innately the same, how should it profit us to work together?

The above 'balanced' argument is rigged, limited to whether 'races' differ, how and why – rather than whether 'race' itself is a meaningful category. Given pupils' exposure to the ideas of inherited 'race' difference, the limited arguments between the two 'eminent' geneticists will fail to challenge those ideas. Pupils will see that one race is on top and the other underneath, and justifiably so! Some have the brain, the others have the brawn; some are born to lead and others to serve. As Fryer points out, it is this kind of thinking – backed up by what purports to be science – that has provided justification for Western economic exploitation of the Third World.

As the Nuffield text admits, 'The great paradox of the problem of racial conflict is that the people who take part in it inevitably call on biology to justify their actions.' This is only too true! By failing to understand this 'paradox' adequately, Nuffield ends up in complicity with it.

Nevertheless, the revised Nuffield text provides some stimulus material and some questions. These could be used as a

Biology Teaching in a Racist Society

starting point in a discussion about 'race' and the crude classification that it involves (see Table 1).

Table 1

What were the features or characteristics that the Nazi regime used to identify the Aryan race?

How accurate or satisfactory, as guides to identification, were these features in practice?

Give examples to support your view.

The data show some of the features of the different races of mankind.

	Number	Complexion	Hair	Stature
Features of different human races	1	dark	dark/wavy	short
	2	fair	dark/straight	short
	3	medium/dark	dark/straight	medium tall
	4	medium/dark	dark/straight	tall
	5	very fair	fair/straight	tall
	6	yellow	dark/straight	short
	7	brown	dark/wavy	tall
	8	yellow/brown	dark/wavy	short

Can you classify the races into groups or subgroups? Show your scheme of classification by joining the numbers (1–8) by lines or brackets. (Nuffield, 1979, p. 20)

Unfortunately, the unit provides no information which would help students to see the concept of 'race' as in itself an ideological fiction, devised only as a means of facilitating genocide and exploitation (Fryer, 1984). Nor does it provide any information which would help students to challenge the notion of 'race' on scientific grounds (Lewontin, 1982). Indeed, the Nuffield text takes the notion of 'race' as given, as if it were scientifically proven fact. Nuffield implicitly suggests that differences in skin colour, body size and hair type represent fundamental and important human differences. Racist ideas are given credence and presented as scientific.

HUMAN VARIATION

None of the biology texts commonly used by schools (indeed, no school text I have read) questions the arbitrary manner in which subdivisions of human groups are made when the text classifies them into 'races'. Why choose hair colour or skin tone? Why not blood groups? In Europe people have differing skin colours, hair types and frequencies of gene for blood group A. Yet somehow Europeans manage to be presented as part of the same race – the whiter and blonder members being slightly more pure, perhaps.

Biology texts fail to examine the variety *between* people of the groups which they designate as races. Nuffield acknowledge this variety, but in racist language (p. 27). As Lewontin points out, there is more variation within a single 'racial' group than between them. There is lots of information concerning other types of discontinuous variation in humans, such as hair type, skin colour and skull shape – all of which are visual. There are also variations which cannot easily be seen, such as haemoglobin structure, ability to taste phenylthiourea, cytochrome sequence, fibrinopeptide sequence and myoglobin sequence, to list a few. In fact, if the frequencies of blood group genes A and B were used as a basis for grouping human beings, then most of Africa, India and China would belong to one group. An anti-racist biology curriculum should include head-on a treatment of 'race' and the racist origins of the concept. We must not be deterred by the excuses of 'conceptually far too difficult', etc., which are often put forward by teachers and others, thus obscuring the issue of what counts as knowledge, and where it comes from.

Biology teachers have to address the question of whether it is desirable to define individual 'races'. Linked to the notion of 'race' is the concept of 'race' purity, which is an essential part of racist ideology. Physical anthropologists have sought to rationalize this by using the terms 'primary' and 'secondary' 'races'.

A 'primary race' is defined as a fundamental extreme and 'secondary race' as the intermediate of two populations. This model is essentially similar to those used to explain the occurrence of various morphs of the peppered moth Biston betularia. Notions of 'racial purity' often lead pupils to ask, 'Have races been pure in the past?' Most physical anthropologists answer this question by arguing that in the past there were fewer people; therefore social groups were more isolated, resulting in greater differentiation among them. Biology teachers would supplement this by showing pupils that the history of people is one of migration and encounter – a fact often neglected in most biology texts.

INTELLIGENCE

As well as tackling the origin of human variation, the biology curriculum needs to help pupils come to terms with arguments about social/genetic intellectual inferiority such as those put forward by Jensen and later by Eysenck. In opposition, teachers can use evidence that shows the effect of environment in regulating the expression of genes. And beyond the mechanistic 'nature versus nurture' model we can show how our society's definition of intelligence itself is culturally specific.

Consider this cultural comparison:

> A group of illiterate Kpelle adults and a group of sixty US Peace Corps volunteers were given two problems. First they were asked to estimate the number of cups of rice that could be obtained from a large bowl, and secondly they were asked to sort in three different ways sets of eight cards, with two or five, red or green squares or triangles on them. In the first experiment the Peace Corps error was on average 35 per cent, four times the average error of the Kpelle. In the second experiment, the Peace Corps volunteers completed the task without hesitation, but the Kpelle found great difficulties, and two-thirds failed to complete three sortings. The point

the investigators make is that to describe the Kpelle performance in the second test as evidence of limited mathematical ability is no more justified than to describe the American volunteers' errors in the first test as inept, unless we already start with preconceptions of mathematical competence.

The general point, therefore, is that if we do not begin by assuming what mathematical (or other) knowledge is, we can compare the way men (*sic*) have constructed – under different social and economic conditions – different styles of thought and kinds of explanation, and how in industrialized societies these have been institutionalized in formal educational systems (Young, 1971, p. 12).

The concept of mathematical ability is seen here to be culturally relative. Other valued abilities can be viewed in the same way. Young disputes the hierarchical definitions of ability which are held by most teachers and institutionalized in the school curriculum and organization. He criticizes conventional sociology of education for accepting these definitions as inviolable, and for failing to recognize how they help to determine what counts as knowledge.

Furthermore, we can examine the way Burt fabricated statistical 'evidence' to support a notion of inherited intellectual characteristics. It would also be useful to consider the influence of Burt's falsified data, beyond the terms of his own work as an educational psychologist. After all, his work was published, widely acclaimed as scientific, and is still to be found, in print and uncriticized, in many a teacher's training college library. Burt was not exposed as a charlatan until 1976. His first paper 'proving' that intellectual differences were innate appeared in 1909; his last was published in 1972, after his death. How many teachers trained before 1976 were brought up on the idea that intelligence may be inherited? Some books published since 1976 still take Burt's arguments seriously. Furthermore, Burt's ideas have been institutionalized in Britain's school system and remain ingrained in popular thinking (See Bibliographical Note in this book's section on Assessment.)

THE TENACITY OF 'RACE'

Ashley Montagu was one of the first anthropologists to argue that the term 'race' be abandoned. His *Statement on Race* (1957) sees culture as a major factor in the development of the human species – 'a unique zone of adaptation', available only to people. He believes that 'race' is far too emotive and subjective a term to be useful to anthropologists. Instead he suggests that we consider humanity as a single species with differing breeding populations, varying in the frequency of one or more genes. To replace 'race' he coined the term 'gene group'.

Montagu was one of the more progressive writers of his time. We must ask why his ideas were largely ignored, while Konrad Lorenz's books enjoyed intensive promotion. Clearly Montagu's ideas did not – and do not – suit those in whose interests the dominant culture is manipulated.

The significance of this issue for schools is enormous. Pupils see differences between people. Physical differences are presented as important *human* differences by popular racist ideology. Political and cultural differences are assumed to be biologically determined. It is intuitively appealing to have one's prejudices and early learning confirmed through formal education.

In an effort to counter racism, Montagu's approach eliminates 'race' as a biological concept. However, 'race' still remains strong as a cultural, political and ideological concept. By virtue of the history of biology as a discipline which has served a racist economic system, physical characteristics are associated with social and cultural characteristics in most people's thinking. Montagu's more progressive approach to 'race' is still inadequate, for his technical approach cannot directly challenge the sources of racism. Teachers need clearly articulated anti-racist strategies which challenge the subject content as well as its traditional boundaries.

In opposition to studying 'racial differences', an anti-racist approach means educating pupils about how they themselves can be involved in the process of transforming the world in which they live. That world is still stamped by the legacy of colonialism and imperialism, which results in such horrors as famine in Africa and 2,500 dead in Bhopal. How can the exploitation of 'developing' (i.e. neocolonial) countries become a thing of the past?

Relevant themes – health, human population and its distribution of diet and nutrition and world resources – are all established areas of the biology curriculum. Regrettably, these are often treated as marginal topics – partly because of the scarcity of resource material, and partly because Britain's past and present involvement in global exploitation make the issues 'sensitive'. Indeed, the topic is regarded as 'political' – whereas its omission, oddly enough, is not generally thought to be so.

The politics of the biology curriculum is a crucial area of analysis for those engaged in a critical appraisal of content and method. What qualifies for inclusion in the curriculum and what is quite deliberately excluded? Teachers need to ask and answer questions like these if they are to take seriously their responsibility in educating for a non-racist society of the future. It is precisely because of the political nature of these questions that the issue of anti-racist education is such a hot potato.

Of course, the content of the curriculum has *never* been politically neutral. Through engaging in anti-racist education, biology teachers begin to realize that their role in society has always been a political one. This can be quite a shock – and a source of shame and guilt, perhaps, for those who have been trained to believe that they are involved in value-free endeavour. Perhaps this is one of the reasons why anti-racist education meets so much resistance.

References

(All books published in London unless otherwise noted.)

Bulldog, 1 September 1977; 2 October 1977.
Stokely Carmichael and Charles Hamilton, *Black Power: The Politics of Liberation in America*, Random, 1967.

Leslie Claydon *et al.*, *Curriculum and Culture: Schooling in a Pluralist Society*, Allen & Unwin, 1978, pp. 59–61.

Peter Fryer, *Staying Power: The History of Black People in Britain*, Pluto, 1984, excerpted in this collection.

Stephen Jay Gould, *The Mismeasure of Man*, Norton, 1981.

Theo Kalikow, 'Konrad Lorenz's "Brown Past"', *J. Hist. Behav. Sci.* 14 (1978), 173–80.

Richard Lewontin, 'Are the Races Different?', *Science for the People*, 14, 2 (Mar./Apr. 1982), 10–14, reprinted in this collection.

Scilla Maclean, ed., *Female Circumcision, Excision and Infibulation*, Minority Rights Group report no. 47, 1980.

Ashley Montagu, *Statement on Race* (1957), Oxford University Press, 1972.

Nuffield Advanced Biology: Organisms and Populations, and *Study Guide*, Longman, 1970.

Nuffield Biology Text 4, *The Perpetuation of Life*, Longman, 1975, revised edition, 1979.

Michael F. D. Young, 'An approach to the Study of Curricula as Socially Organized Knowledge', in M. F. D. Young, *Knowledge and Control: New Directions in the Sociology of Education*, Collier-Macmillan, 1971.

MULTICULTURAL VERSUS ANTI-RACIST SCIENCE: BIOLOGY

DAWN GILL, EUROPE SINGH and MICHAEL VANCE

WHAT IS ANTI-RACIST EDUCATION?

In order to begin to address this question, it is necessary to be clear about what constitutes racism. The term defies rhetorical and easy definition. British racism has its roots in colonialism and imperialism, and may find expression in the social and economic relationship between black and white individuals, in black people's position and treatment within state institutions, and in the relationship of black people (and countries) to the national and global economic institutions; examples are the police, the courts, transnational corporations and the World Bank.

Racism is an integral element in a total social and economic system and in all its institutions. Racism and imperialism are interrelated, as are racism and class, and racism and sexism. Racist ideology and practice are part of the functioning of the capitalist economy locally, nationally and globally. This is not to deny that racism exists in state capitalist societies, nor is it to deny that discrimination is practised in class-stratified black societies, or that black people may engage in exploitative relationships vis-à-vis each other. It is simply to say that discussions of racism cannot be reduced to whether black and white people get on well together when they are in the same room or street. Anti-racism is not just about helping individuals to like each other. It requires a political analysis of an economic system and its historical roots, and of the way economic and ideological systems interrelate.

Anti-racist education involves:
- recognizing racist ideology and practice in all its manifestations;
- engaging the education system (and its personnel) in a recognition of the part it plays in a total ideological and political system which is racist in its operation;
- working out strategies by which the education system can be used to challenge racism in all its manifestations, and thereby become part of the means by which people's relationships to each other and the economy can be altered.

It is impossible to understand racism without a class analysis, just as it is impossible to understand it without a clear perception of imperialism in its historical and contemporary forms. Interpersonal racism must be located within its political, economic and historical context. If racism is to be understood, we need to examine the historical and economic roots of racism, sexism, nationalism and the class structure, and to analyse the ways in which oppressive ideologies are mutually reinforcing. This is by no means a simple task. Educationalists are only now beginning to develop the conceptual tools needed. (For a theoretical framework, see Gill and Singh, 1987.)

TERMS OF THE DEBATE

The terms of the debate on multicultural education were set by the publication of the Interim Report of the Committee of Inquiry into the Education of Children from Ethnic Minority Groups (commonly known as the 'Rampton Report'). As its title suggests, the focus of attention in this debate was 'ethnic minority' children, and more specifically their 'underachievement'. Those children said to be 'underachieving' most dramatically were those of 'West Indian origin'. Many statistics were presented to prove it.

The Committee could identify 'no single cause' of underachievement ... but 'rather a network of widely differing attitudes and expectations on the part of teachers and the education system as a whole, and on the part of West Indian parents which lead the West Indian child to have particular difficulties and face particular hurdles in achieving his or her full potential.' The net result of the furore caused by the publication of this report was that black pupils themselves – and their families – became the objects of scrutiny.

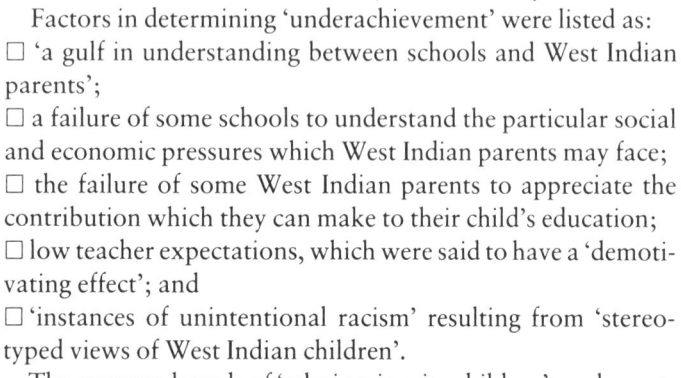

Factors in determining 'underachievement' were listed as:
☐ 'a gulf in understanding between schools and West Indian parents';
☐ a failure of some schools to understand the particular social and economic pressures which West Indian parents may face;
☐ the failure of some West Indian parents to appreciate the contribution which they can make to their child's education;
☐ low teacher expectations, which were said to have a 'demotivating effect'; and
☐ 'instances of unintentional racism' resulting from 'stereotyped views of West Indian children'.

The supposed needs of 'ethnic minority children' can be met, the report suggested, through a programme of 'multicultural education'. Since the mid-1970s multicultural education has also been seen as a way of reducing white people's hostility towards minority groups. Multiculturalism assumes that if pupils learn to appreciate each other's cultures, then all will be well. The celebration of diversity is seen as a good thing, because it will lead to harmonious relationships between groups of people who may be at present antagonistic towards each other. (For this perspective, see Willey, 1982.)

These simplistic premises underlie the Rampton Report, which summarizes the broad aims of a multicultural curriculum as trying to ensure that:

> All children learn about their own cultures and histories and see them treated with equal respect.
>
> All children are equipped with the necessary skills and information to have access to the culture of their own communication and of other countries.
>
> All children fully appreciate the important contributions which ethnic minorities make to this society.

The knowledge and values transmitted by the school seek to remove the ignorance upon which much prejudice and discrimination is based.

Positive attitudes towards cultural diversity are developed so that society can build on and benefit from the strengths and richness it brings. (DES, 1981)

'Multicultural education', and those who promote it, make a number of naive assumptions:
☐ that culture is an easily understood and identifiable social phenomenon specific to each 'ethnic' group;
☐ that cultural oppression is the fundamental problem facing black people and black children in schools;
☐ a positive image of black people and their cultures would have a positive influence on pupil motivation and school performance;
☐ that we have to learn tolerance, which means accepting all cultures as equally valid, without criticism; and
☐ that this tolerance and the knowledge of 'other cultures' break down racism amongst white students.

These assumptions are all without foundation (Gill and Singh, 1987). For example, the 'common-sense' racism of the white working class – which multiculturalist strategies aim to overcome – is not based on cultural ignorance at all. If anything, it is the result of ignorance about the economic and political causes and consequences of immigration. With such ignorance it can seem obvious to blame 'immigrants' for 'taking our houses and jobs' (Sivanandan, 1978). For many years the mass media and political parties have fostered such attitudes – which multicultural education does little to challenge.

MULTICULTURAL PRACTICE

Multicultural education impinges minimally on the mainstream curriculum; instead it concerns itself with the perceived 'special needs' of pupils from ethnic minority backgrounds.

128 Anti-Racist Science Teaching

English as a Second Language receives a place on the timetable; in some enlightened schools there are mother-tongue teachers; school assemblies may celebrate the festivals of religions other than Christianity; and home economics classes experiment with curries. There may be lessons about 'other' people and 'their' cultures for the white British pupils; usually these lessons form part of social or religious education – typically, marginalized curriculum areas with no examination status. A study of other people and sterotyped versions of their cultures is often interpreted in terms of the clothes they wear, their art, the food they eat, their religions and their music.

Where there is 'multicultural' input into the science curriculum it tends to focus on so-called 'Third World Science' and involves activities like making salt from banana skins (Watts, 1984). The patronizing view of the 'clever and resourceful native' which underlies such practice is not far removed from the racist views of 'other people and their cultures' which pervade attempts at multicultural education in geography.

Multicultural education – whether in geography, history or science – is essentially patronizing and voyeuristic. It fails to place 'other cultures' in their political and economic context. This failure is exemplified by a collection of science teaching materials recently published from a unit within the Inner London Education Authority (NLSC, 1986). It uses the term 'multicultural' interchangeably with 'anti-racist', reduces racism to personal prejudice and reduces anti-racism to valuing cultural differences. Only one section of the book (on copper) even hints at ways of seeing the roots of racism in economic exploitation, or how Western science has expropriated the science of other cultures and turned it against them.

The next section outlines ways in which an anti-racist approach differs from a multicultural one.

ANTI-RACIST OR MULTICULTURAL?

Good biology is necessarily anti-racist biology, emphasizing political, historical, social and economic aspects. These should be reflected in the assessment procedure of any biology curriculum.

It should aim:

☐ To expose and combat the way that racist, sexist and imperialist ideology is mediated through traditional curricula and textbooks.

☐ To provide alternative perspectives to the Western capitalist world-view which currently predominates in teaching, especially regarding poverty and 'Third World underdevelopment'.

☐ To support black and 'ethnic minority' pupils whose confidence and self-esteem have been eroded by the content of biology teaching in the past.

☐ To draw parallels between racism, sexism, and discriminatory practices based on social class: in this context, to foster solidarity between boys and girls and black and white working-class students by challenging the uses of IQ and other normative testing in relation to 'race', class and gender.

☐ To expose the concept of 'race' as a fiction devoid of scientific validity, though having a deep ideological purpose.

There has been no generally accepted change in the secondary school biology curriculum which promotes anti-racist education. However, there is some innovation at the level of policy, as well as the initiatives and practice of some biology teachers, although there is evidence of confused thinking on the distinction between multiculturalism and anti-racist education.

'Good Practice'?

In 1972 Townsend and Brittan tested the ground for curriculum change when they surveyed schools and their science departments. Twenty heads of science departments listed six topics within the scope of biology as areas which prepared pupils for life in a multicultural society. These six areas were: genetics, evolution, blood groups, skin colour, adaptations to environment, food and nutrition. A few departments suggested greater emphasis on human biology, physiology, heredity and health education. Nevertheless, teaching these topics in the absence of a coherent political and economic framework is likely to promote racist thinking – more by default than by design, perhaps, but good intentions are no excuse.

Willey (1982) describes some approaches which he suggests can be taken up by any biology department which is reviewing its courses:

> 1: The curriculum should aim to promote an awareness of racial differences and their origins, and to explain these differences in terms of biological principle of variation, evolution, natural selection and adaptation to environment.
> 2. By avoiding racial stereotyping and offence regarding religious or other observances/beliefs.
> 3. Taking care in choice of illustrations of physiological and biochemical phenomena.
> 4. By using examples from minority as well as the majority groups in explanations and descriptions of society and man's [sic] development.
>
> Where we deal with matters which affect race [sic] we take a broad *multicultural* view and attempt to avoid a negative approach when dealing with biological and environmental problems associated with Third World countries, and particularly the countries of origin of our own pupils.

Willey's idea that we can take a broad *multicultural* view of *race* indicates precisely the kind of confusion which dogs any discussion on multicultural education. The very idea of race is an ideological fiction which has a political function. As the article by Fryer makes clear, the term was pressed into service in the eighteenth century to 'justify' genocide and economic exploitation. Instead of taking a 'multicultural' view of a term which is bankrupt of scientific meaning, we should be questioning the validity of the term itself and identifying its political and ideological significance.

Willey's suggestion that biological and environmental problems are 'associated with' Third World countries is another collection of words which misrepresents the economic and political position of the so-called Third World in the global economy. These 'problems' 'associated with' the Third World are largely a result of an ongoing *process* of underdevelopment. Willey's representation of the situation does not carry a hint of this.

Further supposed 'good practice' is mentioned by an HMI report (1979):

> In four of the schools, work in science was seen which could be described as cross-cultural. An Asian teacher of science described the food he ate at home and compared its content and food value with comparable Western dishes. Another teacher outlined the science of pigmentation using different pupils in his class as a living resource.
>
> In yet another school genetics was being taught with sensitive reference to current theories about racial types and their supposed characteristics.

Both the HMI Report and Willey's point to biology as the science area which most readily lends itself to educating students for a multicultural society. Unfortunately, neither publication has moved on from the approach endorsed by Townsend and Brittan ten years earlier. The Third World is still seen as fraught with its own problems, while its relationship to the 'developed' West remains unexamined. The patronizing, naive and essentially racist nature of these official guidelines illustrates the danger of an uncritical multicultural perspective; it highlights the need for re-educating the HMI and others about racism in society and racist ideology in the curriculum.

Survey of Practice
Recent research undertaken for the Schools Council (Vance, 1984) allowed an appraisal of the current practice concerning a biology curriculum for Britain's multicultural society. Thirty-nine school biology departments responded to a survey of 'Mode 3' syllabuses. None of the respondents referred to the multicultural nature of their school as a major reason for developing such biology examinations, although 'the needs of their pupils' and 'improving exam results' were cited as typical reasons.

132 Anti-Racist Science Teaching

It is sobering to note that none of the respondents mentioned racism as a problem with which the biology curriculum should concern itself. Some respondents expressed a willingness to adopt a 'multicultural perspective', but added that resource material was scarce. One wrote about a discussion that had taken place at her school within the department: as a black British teacher, she felt that the most important task was to get her colleagues to be aware enough to draw out the particular cultural experiences of pupils during question and answer sessions, as well as using examples from other cultures. The department has made a start by widening the scope of their practical work in the third year; it includes looking at a wide range of foods, examining different styles of hair and skin care, and considering the types of pupils in the class, using *Science at Work – Cosmetics* as a guide (Taylor, 1979).

No doubt the girls were comparatively well behaved when presented with something as dear to their hearts as a study of cosmetics: it is fun to be socialized into an inferior economic position where you are expected to take a decorative and servile role. Can we, however, condone this in the interests of promoting harmony in a multicultural society? And what has it to do with racism? Does this approach challenge European standards of beauty? A focus on food, cooking and hairstyles does nothing to help pupils challenge racism – and still less in terms of challenging sexism. It may be well-meaning, but it is essentially quite patronizing and diversionary: it reflects the latest variant of oppressive pedagogy and content.

Respondents listed the following topics, already included in their biology 16+ syllabuses, as being of relevance to a 'multicultural' approach to the subject:

☐ energy requirements and actual energy intake in different countries;
☐ testing a wide variety of foods – including yam, sweet potato, pasta, green peppers – for starch, fat, sugar, protein and vitamin C content;
☐ oral hygiene and the incidence of dental caries in relation to diet;

☐ hay fever and its incidence amongst 'ethnic' minorities;
☐ origin of blood groups;
☐ the role of melanin in the skin, carotene;
☐ diseases of worldwide significance;
☐ how we taste spicy foods;
☐ the genetics of sickle cell anaemia;
☐ the distribution of races; and
☐ racial morphology.

The respondents also considered which topics of a multicultural nature might be included in future Mode 3 courses. Some of those included overlap to a certain extent with the list above:

☐ sickle cell anaemia, frequency and heredity;
☐ blood groups and race;
☐ heart disease and diet;
☐ world diets, their energy values, their significance; comparison with UK diet;
☐ food tests to incorporate foods of cultural minorities;
☐ diet deficiency diseases treated alongside obesity;
☐ famine and starvation, world food resources and inequalities;
☐ diseases of the developed and underdeveloped world;
☐ racial variation and adaptation, evolution of races;
☐ pigmentation, its significance; and
☐ comparative medicine and hygiene.

Two respondents suggested project work as a vehicle for bringing 'multiculturalism' into the mainstream biology examination curriculum. However, a large part of what is given the seal of approval by 'multiculturalists' is highly suspect from an anti-racist standpoint. Why focus on skin colour, for example, as a significant and important area of study? In our racist society skin colour has been ascribed an importance out of all proportion to its biological significance. Its real importance is political. To examine skin colour as something *biologically* important is racist, unless it is done as part of analysing racism as a social phenomenon. 'Race' in itself – and skin colour – are meaningless outside that context. Indeed, studies of the distribution of supposed 'racial' morphology and the evaluation of 'races' are clearly racist in that such studies presuppose the existence of separate races.

Poverty in Multiculturalism
A study of famine and starvation in today's world is all well and good – if it is analysed within a framework which relates starvation to the production and distribution of wealth at local and global scales. Starvation is not suffered by the rich; it is suffered by the poor. Poverty is neither natural nor inevitable. How is it created? What is its relationship to wealth? Present biology courses focus on the symptoms of poverty – kwashiorkor, rickets, cholera, tuberculosis, malnutrition – while neglecting its causes. Explanations of poverty frequently locate its cause within the individuals or nations suffering the problem, not within an economic and political system which forces them to do so.

The curriculum as it currently operates serves to isolate one area of understanding from others. Students who learn about diseases and starvation in biology are unlikely to do so in a way which enables them to understand the political and economic roots of these problems. Thus it happens that individuals in the 'Third World' may be held responsible for the poverty in which they live. The content of a 'multicultural' biology curriculum coheres with the confused and dishonest explanations of poverty to be found in geography and history curricula. The whole package results in racist interpretations of global political and economic issues (CIGE, 1984).

Unfortunately, the scope of the Schools Council research did not allow examination of the curriculum in sufficient depth to determine what was actually happening in classrooms, where 'multicultural education' was supposed to be taking place. There is clearly a need for further research in this area. It would have been interesting to research more deeply into what, exactly, was taught in schools: in dealing with poverty-related health problems, for example, do teachers neglect the causes of poverty, while focusing on its consequences?

Such topics can easily be covered in a way that fosters Western preconceptions. Unless these issues are explored within a political and economic framework they are likely to promote not only racist attitudes, but racism in its broader sense.

References

(All books published in London unless otherwise noted.)

Contemporary Issues in Geography and Education, 1, 3 (1984): 'The Global Economy: Trade, Aid and Multinationals'.
DES, *West Indian Children in Our Schools* (Rampton Report), HMSO, 1981.
D. Gill and E. Singh, *A Beginner's Guide to Anti-Racist Education*, ACD, 1987.
D. Gill and E. Singh, *'No Racism Here'*, ACD, 1987.
HMI, *Aspects of Organization and Curriculum in Seven Multi-Ethnic Comprehensive Schools*, DES/HMSO, 1979.
D. Mackean and B. Jones, *Human and Social Biology*, John Murray, 1975.
North London Science Centre, *Science Teaching in a Multi-Ethnic Society*, Book Two, 1986.
Walter Rodney, *How Europe Underdeveloped Africa*, Bogle L'Ouverture, 1972.
A. Sivanandan, 'From Immigration Control to "Induced Repatriation"', *Race & Class*, XX, 1 (Summer 1978); reprinted as Race and Class pamphlet no. 5.
John Taylor, Project Director, *Cosmetics*, Science at Work series, Addison Wesley, 1979.
H. Townsend and E. Brittan, *Multiracial Education – Need and Innovation*, Schools Council Working Paper 50, 1973.
M. Vance, *Assessment in a Multicultural Society: Biology at 16+*, Longman for the Schools Council, 1984.
P. Whitten and J. Kagan, 'Race and IQ', in D. E. Hunter and P. Whitten, eds, *Readings in Physical Anthropology and Archeology*, NY, Harper & Row, 1978.
R. Willey, *Teaching in Multicultural Britain*, Longmans Resources Unit for the Schools Council, 1982.

KENYA: THE CONSERVATIONISTS' BLUNDER

MALCOLM GREEN

Based on a Treatise by Walter Jamie Lusigi, *Planning Human Activities on Protected Natural Ecosystems*, Part One, published in West Germany by J. Cramer, 1978.

The images we receive of Africa in this country are usually negative ones: of natural disasters, war and poverty. Positive images are often of wildlife, wide open spaces and physical beauty; when people enter into these latter scenes they are likely to be 'primitives' in tribal ritual, cruel hunters or helpless people. There are exceptions to these images, especially now that television is producing more enlightened programmes. We are made aware that the wildlife of Africa is threatened, partly by the import of skins and ivory to the West. Yet most of us still assume that the wildlife in such countries is endangered basically by human greed and expanding populations. In the minds of English children, many a fantasy world must have been built up around the vast plains of the Serengeti. We are told *we* must help 'save' the remaining white rhinos or cheetahs, without considering how our method might affect the African people.

W. J. Lusigi, an ecologist working for UNESCO, has written a critique of that approach. In his Part One, on which I base this essay, he looks at the way European influence has led to a deteriorating relationship between the African people and wildlife. In particular he examines how Western ideologies of conservation, imposed upon the people, have contributed to this deterioration in Kenya.

Lusigi explains how the European attitude to wildlife and conservation in Kenya – much of it founded on alien philosophies – is responsible for many of today's problems. He writes, 'At present conservation in Kenya is an outgrowth of Western civilization's needs, fears and values rather than concerned with those whose lives are immediately affected.' The threat to the existence of wildlife in Kenya comes from a far more complex human motive than simply greed or pressure for land. Most Kenyans 'do not find the idea of wildlife conservation either intellectually or emotionally satisfying . . . and do not support it.' It is often associated with misery and oppression. Some people are actively antagonistic towards it.

Lusigi's research ascertained the reason for these feelings, given that people and wildlife coexisted in plenty up until comparatively recent times. He interviewed old people from a number of different tribes. As far as he could, he traced people's attitudes back through history.

To understand the cultural differences involved, let us consider today's Western approaches to conservation. For example, Britain has a vast array of different schemes designed to protect parcels of our fast-disappearing countryside. They all have basically one thing in common: to minimize human interference, except where management is required, in order to ensure the perpetuation of a natural or semi-natural ecosystem and its contingent wildlife. Normal human activity – especially modern agriculture – is rarely seen as compatible with diverse natural ecosystems; neither is wildlife regarded as necessary for human survival.

When the first European explorers went to Africa and witnessed the direct relationship with and use of wildlife by African people, they labelled them primitive, savage and uncivilized. Europeans used this behaviour as the foundation for many racist attitudes that emerged. At the time, there was little interest in the social subtleties and complexities that enabled these people to survive in very harsh conditions. No account was taken of these when 'development' was forced upon Kenya, with subsequent disastrous consequences for both the people and the environment. There is little doubt that the terrible famines striking Africa today are, in part, a consequence of destroying the balance and safeguards that the people had built up in their relationship with nature. The British attitude to wildlife conservation played its role in this. The colonial administrators sought to safeguard Kenyan wildlife without any consultation with the African communities that coexisted with them – an amazing arrogance when one considers how long the Africans had been custodians of the game before the Europeans set foot on the continent.

Lusigi talks of three periods in history that formed present-day attitudes of Kenyans to wildlife conservation: the ivory trade; European settlement and colonialism; independence.

THE IVORY TRADE

Although the history of ivory trading in Kenya is recorded as far back as 110 AD, the critical events happened in its last fifty years. Africans in the interior (especially the Akamba) collected ivory and transported it to the coasts where, at different stages in history, it was bought by Chinese, Indians and Europeans for shipping abroad. There was much intertribal trade, and from the sixteenth to the nineteenth century the people of the interior of Kenya flourished. As Burton described the African's life then, 'his condition may be compared with the peasantry of the richest parts of Europe.' At the same time, large towns grew up on the coasts.

However, slavery grew out of this ivory trade and gradually increased, reaching a climax during European trade from the 1880s to 1890s. Much suffering was inflicted on the people of the interior, who were forced to carry heavy loads of ivory on their heads for mile after mile. Pressure was exacerbated when the Europeans traded guns in exchange; conflicts immediately started to flare up. Even with abolition and the cessation of slave trading at the coast, slave labour increased in the interior, as it undercut the price of porters. By the early 1900s there was widespread famine. Wissmann records devastation and misery in previously beautiful villages. When Lusigi interviewed the Akamba people about this period, they sang songs of misery.

Kenya: The Conservationists' Blunder 139

Wildlife was seen as the source of much of this suffering and there was a polarization in people's attitudes.

At about this time the colonial administrators began to realize the devastation that had ravaged the stocks of wildlife, in particular the elephant, which many thought could soon become extinct. It would have been the ideal time to call on the African pastoralists' intimate knowledge of the indigenous wildlife, but instead they were blamed for its demise and treated as if they were inimical to it. The ignorant European attitude is illustrated by the words of Sir Charles Elliott: 'We are not destroying any old or interesting system, but introducing order into a blank, uninteresting, brutal barbarism.' The more enlightened Wissmann observed the way Europeans careered around the countryside blasting at every living creature. In line with recent conservation developments in the USA, restrictions started to be put on wildlife hunting. Those most affected – Africans – were not consulted. Even worse, the restrictions came at a time of drought, when most people were most dependent on game for food. The Africans found the new laws incomprehensible and cruel.

The restrictions took the form of a succession of different conservation measures – hunting licences, game reserves, and, most recently, the National Parks. These gave wildlife varying degrees of protection from people. The National Park doctrine excluded people totally from protected areas, except as passive sightseers. There was scant regard for the role which local tribes had played over the centuries in the formation of the spectacular game habitat. 'The consequence of these restrictions was devastating for a number of tribal groups.' The Waliangula, who depended on elephants for food, found three-fifths (406) of their male population declared poachers and imprisoned. The tribal society totally disintegrated. On their release from prison, many such 'criminals' resorted to petty crime in the cities, whilst the few that returned to their villages became bitter and forlorn. Similar incidents affected many groups.

Whilst game laws were being imposed, white settlers were colonizing the plains of Kenya, eliminating the vast herds of herbivores that competed with their cattle. Seemingly empty of other human activity, these lands were in fact crucial to the survival strategies of the Masai and other pastoralists, who depended on them for grazing in times of drought. Settlement by the Whites reduced their options and placed an intolerable stress on the people, resulting in conflicts and sporadic war between different groups and making famine ever more likely. (Ironically, the West today sees famines as 'natural disasters' that the people are helpless to deal with.) The independent-spirited Masai, whose freedom to move has been more and more restricted, are now a dispirited people who tend to see wildlife as a major factor depriving them of their lands for their own traditional development.

The First and Second World Wars both had repercussions for wildlife and people's attitudes to it, when Africans were ordered to shoot vast numbers of game to feed the armies and prisoners of North Africa.

INDEPENDENCE

Independence was a third phase in the polarization of Kenyans' attitude to wildlife:

> When Kenya got independence there were many problems to be overcome in adjusting the running of the country to the needs of the people. Independence meant different things to different people . . . to the nomadic population it meant the end to all restrictions and game laws, which were seen as the colonialists' ways of keeping the Africans down.

However, people were all united in the wish to see Africans in the positions previously held by Europeans. The Europeans had deliberately not done a good job in training Africans in preparation for this takeover, especially in fields like wildlife management, where there was not a single African specialist. This made impossible what was really needed – a radical new perspective on wildlife conservation to bring it back into the Kenyan culture. With no decisive initiative to put right the problems created during colonialism, people in the wildlife

Kenya: The Conservationists' Blunder

areas became very disappointed and the situation deteriorated further.

At the time, wildlife conservation was split between the Game Department and the National Parks. Although the Game Department was Africanized, the National Parks were a quasi-governmental body, remaining basically in the hands of Europeans. The Parks threw up administrative walls to shield themselves from the unstable conditions during transition and attracted large amounts of overseas aid. Although this brought some internal benefits, it further isolated conservation from the Kenyan people and brought about considerable antagonism. It was not until 1976 that the National Parks were incorporated into the Kenyan government.

Since then the Kenyan government has consistently faced a dilemma over these National Parks: the question of land ownership. In the USA, when Indians claim territorial rights over National Parks land, these are easily overridden. Kenya,

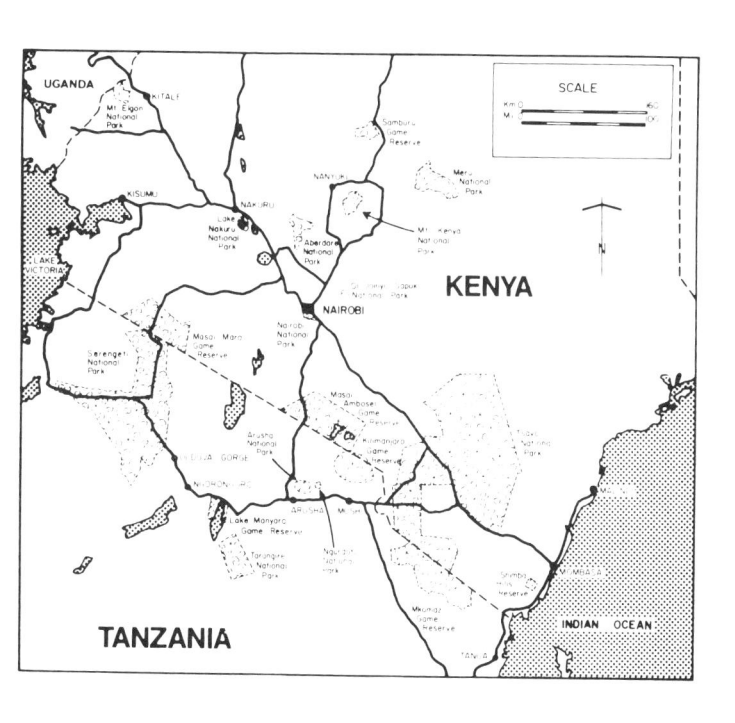

GAME LAWS FOR TOURISM

Tourism in East Africa developed out of the interest of Europeans and Americans in hunting and seeing big game. Such an interest resulted in the establishment of game laws and eventually reserves or parks that have helped to ensure that such a tourist attraction is available today.

Kenya was the first country in Africa to establish laws relating to wildlife, these being the East Africa Game Regulations of 1900, which were refined as The Game Ordinance, 1908. In 1909, the protectorate proclaimed the Southern Reserve (in Masailand) and a Northern Reserve. In the 1920s and 1930s there was pressure, mainly from the colonial settlers and foreigners, to establish national parks, with the Kruger National Park as a model.

An international conference on wildlife in 1933 in London reached a general agreement on the principle of establishing and maintaining national parks or permanent faunal sanctuaries. The London *Times* that year observed with respect to East Africa that 'the value of proper parks, from the economic and tourist point of view, needs no elaboration.' It is important to note the almost totally external pressure for, and organization of, parks at this time – a precedent which lasts to the present. However, a 1937 Ordinance left Kenya with only two game reserves covering 20,000 square miles. Finally in 1946 the Nairobi National Park was established, then in 1948 Tsavo National Park, 1949 Mt Kenya, 1950 the Aberdares, and in 1967 Marsabit and Mt Elgon. (See Map.)

In Tanganyika (now Tanzania) the German colonizers also exerted external pressure to develop game laws and parks. In 1912 a long list of reserves, including many areas which are now parks, was defined. The British maintained these laws when they acquired the territory in 1917. Tanganyika led East Africa in establishing national parks by proclaiming the Serengeti Plains a park in 1941. The area was expected to 'provide an attraction to tourists not to be found elsewhere in Africa', and the presence of the Masai was deemed not 'incompatible with the purposes of the park'. However, this again was an external viewpoint; the problem of the cultural impact of tourist facilities had begun. In 1959 the boundaries of the park were revised, leaving the Ngorongoro Crater area for the use of the Masai.

Uganda's game laws go back to 1906, with game reserves being established shortly thereafter. The Murchison Falls and Queen Elizabeth National Parks, based on earlier reserves, were established in 1952. Their names again reflect the external pressure for, and tourist appeal of, such areas.

Early tourism in East Africa was based on the availability of game for hunting and its protection by law. Until the post-World War II period the tourist was a rarity, an affluent foreigner able to afford weeks or months (including the long period of travel to and from East Africa) to enjoy a safari. Such activity has been described, and usually glorified, by such writers as Hemingway, Roosevelt and Hunter. The long gestation period of East African tourism helped develop a strong image and pattern of tourist behaviour that leaves its mark today.

Excerpt from John S. Marsh, 'Tourism and Development: The East African Case', *Alternatives* 5, 1 (1975), 15–21, Peterborough, Ontario.)

however, has an African government responsible to African people, so ancient tribal land rights are not so easily ignored – yet they are still denied. To enter a park today, one must be in a car and one must pay – options not open to many rural Kenyans. Administrative buildings, accommodation for staff and all tourist facilities are housed in the parks, very often near water and places important for animals. Not only does this strain ecosystems, but in addition it creates isolated islands of prejudice for the white tourists who are allowed to occupy the land, previously the territory of those on the other side of the fence.

RETURN TO TRADITION

Thus National Parks and wildlife conservation in Kenya have served the interests of whites, from their origins to the present time. The Africans see wildlife as responsible for depriving them of their lands and rights. As the parks cannot long survive in their present form, Lusigi looks towards African tradition to find a solution. He says: 'I have fallen back on tradition and culture because it is still the major dominating force in our society . . . the era of colonialism was too short . . . to destroy or significantly depreciate the essential elements of the indigenous culture and tradition.'

Lusigi talks at some length about aspects of the African traditional culture relevant to wildlife conservation; especially spiritual beliefs, land ownership, hunting and education. He sums up by saying:

> Through long traditional living they had evolved a balance between them and the ecosystem, in fact they were part of it. They depended on the ecosystem for their survival. They knew that if they persistently violated ecological rules they would perish . . . this balance has been destroyed by Western conservation ideology, poaching, and commercial exploitation of the wildlife resource.

'Wise use was the rule of African resource exploitation,' he said, and 'they believed strongly in the passing on of these

Kenya: The Conservationists' Blunder 145

resources to future generations.' 'Use' is the key word here. The idea of 'preservation', on which early wildlife conservation was based, is not in tune with it. 'We will never have a landscape again teeming with wildlife.' To be consistent with African thinking, 'it is virtually necessary to utilize the wild fauna in order to ensure their survival.'

Lusigi brings out three principles from African tradition as a vital philosophical base for wildlife policy today:
1. Wildlife is 'common' and to be used for the benefit of all members of the community.
2. Wildlife is culturally God-given, passed on by the ancestors for our survival. It is our responsibility to pass it on to our offspring.
3. Although all individuals have rights to the land, it is a 'common' which must be used in such a way that it can provide a home for us and wildlife for the present and future generations.

Practically speaking, Lusigi sees the solution in bringing National Parks back into regional land-use plans, so that they are integrated economically and ecologically. He sees the National Park as the centre in a mosaic of land uses. It provides total protection for wildlife in the centre, whilst outside there are consecutive rings with increasing human activity. At present, hunting is totally banned in Kenya; Lusigi suggests instead that controlled hunting be again allowed in areas outside the park for meat purposes. Lusigi's other suggestions all focus around increased African involvement in the parks, with free entrance to locals and consultation with those whose land it is by tradition. He suggests that tourist centres be developed outside park boundaries and local Africans given the key administrative roles.

Basically, he is trying to challenge conservationists' idea of protecting wildlife in segregated segments by bringing wildlife areas back in line with the Africans' idea of being part of a form of land use. Wildlife would then cease to be the domain of the privileged few – mostly white tourists who have the leisure to afford it – and might gradually become reincorporated as a resource for Kenyan rural culture. Lusigi sees this as the only practical solution for ensuring the long-term survival of Kenyan wildlife. (The alternative would be stiffer controls against the Africans, whose resentment would lead them to breach the fences excluding them.)

Perhaps there is some naivety on Lusigi's part in drawing upon Kenyan traditional values within Kenya's present-day competitive market economy, in which wildlife inevitably becomes a commodity. To avoid that, in Part Two of his critique he makes detailed practical proposals for Nairobi National Park: a large core area would remain free from hunting and tourist traffic, while surrounding zones would allow human intervention. Rural Africans would take responsibility for the wildlife; they would benefit from it both by traditional uses and by the tourist revenue it attracts. In this way Kenyan traditional values, recognizing the intrinsic value of wildlife to people, might prevail over the market values of modern capitalist Kenya.

SCIENCE CURRICULUM INNOVATION AT HOLLAND PARK SCHOOL

DAWN GILL, VINOD PATEL, ASHOK SETHI
and HENRY SMITH

The 'comprehensivization' of education has been carried out in the name of equal opportunity for all students, regardless of their sex, skin colour, class, religion or cultural background. Secondary education in London became officially comprehensive during the late 1960s and early 1970s. However, we are still a long way from the time when teachers and administrators can claim that the school system provides equality of opportunity.

Inequalities in educational provision have many manifestations and many causes. These must all be analysed and addressed in the curriculum and organization of schools if schooling is to play any part in challenging inequalities. Of course, the school system cannot solve the problems generated by the social structure which it exists to serve. It is important to recognize what schooling cannot achieve, while at the same time recognizing educators' potential for helping people to learn to take control over their own lives. As teachers, we must not delude ourselves into thinking that we can do any more than educate – but it is in the struggle to do this honestly and fairly that we are faced with enormous issues of inequality and all its manifestations.

Educational policies cannot be divorced from policies of health, housing, employment: there are no purely educational problems. However, education can play a part in helping students to understand the forms of oppression – such as racism, sexism and social class – that affect their lives. The science staff at Holland Park School, after lengthy discussion, decided that the content of education and the organization of learning environments needed to be changed if the Education Authority's equal opportunities policies were to be translated into practice through science education.

Pupil and teacher expectations and relationships in Britain have been conditioned by traditional models of education. This is a barrier to curriculum change, as are factors such as the organization of the school day and the concept of knowledge as a series of separable 'subjects'. However, progress is not impossible, although it may be slower than anticipated. The determined pursuit of clearly articulated objectives by a committed staff can achieve significant results, even within the constraints imposed by examination syllabuses and existing preconceptions of education.

We cannot claim total success in introducing curriculum innovation, but we have learned a lot from our experience. Others may value the opportunity to see what we attempted to do, and why, and what went wrong, and why. Our efforts have met with some success; teachers in other schools may be interested to learn how this was achieved.

What follows is a description of our school's effort to translate equal opportunities rhetoric into reality, through curriculum change in science. We also discuss the difficulties of introducing innovative content and pedagogy.

THE EARLY YEARS OF COMPREHENSIVE EDUCATION

Holland Park has been nominally a 'comprehensive' secondary school since it opened in 1958. Since that date most staff have been working together in an attempt to make comprehensive education as much of a reality as possible in the school.

Mixed-ability classes have been in operation in the lower school since 1971. Where there were comparable GCE and CSE syllabuses, upper-school classes (fourth and fifth year) were also 'mixed-ability'. Such was the commitment to mixed-ability education that exam results in 'mixed' classes were as good as, or even better than, in streamed classes.

Maths, English and science had been compulsory until the

16+ stage since 1979, but options were allowed with the sciences chosen. Even with all the 'advice' offered, the mini-option allowed within the science area tended to lead to the traditional stereotype choices: Craft, Design and Technology (then termed woodwork/metalwork/engineering/motor vehicle maintenance) by the so-called 'underachievers' – good with their hands!; home economics by girls; electronics by black boys; biology by girls; physics by boys, etc.

The science staff were aware that quite often the only result of our teaching effort was to 'qualify' a very small percentage of students to go on to A levels and then college. The 'neutral' science concepts of acceleration, preparation of oxygen or photosynthesis, etc., seemed neither to make sense to the vast majority of our students nor to connect to their beliefs in any way. And yet we knew that this majority of students were not anti-scientific, illiterate or uninterested.

The Science Department aims of 1976, which reflect our development in the mid-1970s, show that teachers had begun to think critically about the traditional content of science education. We wanted all students to study the social, political and economic bases of science as part of their lessons.

HOLLAND PARK SCHOOL SCIENCE DEPARTMENT

Aims (1976)
1. Acquisition of experimental technique and quantifying technique and experience of these.
2. Development of linguistic and numerical skills.
3. Development of the students' natural curiosity and critical analysis with conclusions. Comprehension of hypothetico-deductive model.
4. Realization of the scope of science.
5. Showing the relevance of science to social, political, economic and environmental factors.
6. Development of the ability to utilize sources of information.
7. To enable students to work confidently both as individuals and as co-operative members of a team.
8. To assess whether we are achieving the above aims.
9. The development of science as an integrated study.

Our aims were never fully achieved, although we did get a little way down the road. After option choice had taken place during the third year, individual teachers went off with separate classes, into separate laboratories doing separate subjects leading to separate exams – all of which we probably did as well as most schools. During this time various attempts at, for example, team teaching or group work had their rationale in providing support for some teachers who found it difficult to cope with the formal nature of post-option exam classes rather than in any desire to make, say, chemistry more inviting an area for students to study.

Science Curriculum Innovation 151

Most of our lessons in the third/fourth/fifth years were very prescribed; syllabuses needed covering in a very organized fashion with well-established routines. Lessons often had to be planned out for half a term ahead to ensure that topics would be covered and that technicians knew what equipment to prepare for practical work. Teacher turnover was quite high in ILEA – given the stressful work and low pay – with staff leaving and joining the school almost every term. It appeared that the 'permanent' staff could get some continuity by supplying a prescribed 'teacher-proof' package of lesson plans which could be used by the floating population of teachers who came and went without stopping for long.

All too often, within the time allowed for examination work, it seemed impossible for most teachers to cover the exam syllabus except by using a formal didactic 'chalk and talk' approach. This meant that students were left behind unless they possessed the skills required, such as note-taking, homework completion, formal language use, individual study, etc. The formal classroom/laboratory seemed to divorce, almost naturally, the chemistry or physics being studied from any life outside the laboratory. Looking back, it now comes as no surprise to find that the working-class/black/female students were the ones disadvantaged by this process.

By the late 1970s it was clear that the school was failing in its quest to provide 'equal opportunities for all'. It had become increasingly obvious that the subject content and methods in science needed critical appraisal. Many of our students were not doing well in relation to the standards set by the exam system, yet we knew that these students were intelligent and capable people. Disadvantage for these pupils was somehow institutionalized within the curriculum and examination system.

AMALGAMATION: COMMON-CORE CURRICULUM

The early 1980s brought an increasing sense of frustration, along with greater determination to work with our students in overcoming the obstacles placed in their way by the examination system and the teaching methods associated with it. At least some of the barriers to 'equal opportunity' could be overcome within the school. We had become particularly critical of subject content and pedagogies. We had begun to work towards new teaching methods as part of our programme of 'mixed-ability' teaching.

In 1982–3 the school was faced with a major reorganization in amalgamating with another two schools. This was an extremely painful and stressful process, involving competition between colleagues for jobs, and potential staff redundancies. However, it did at least provide the impetus for a total review of the school and its work. All courses, methods of classroom organization and styles of teaching and learning were subjected to critical evaluation. The 1984 aims of the CDTMS faculty, the result of this evaluation, reflect our thinking in the early 1980s.

At amalgamation we established a common-curriculum approach as being more in line with a comprehensive education offering equal opportunity. If an educational experience was worth having, then surely all our students deserved that opportunity. The school staff decided to ease the introduction of the common curriculum by introducing a faculty structure. This would help staff communication and the co-ordination of their work, and eliminate overlap between subject contents.

'CDTMS' faculty (Craft, Design and Technology, Maths and Science) was formed; this included the former Home Economics Department as Food Technology and Fabric Technology. All students aged eleven to sixteen were to study within this area.

AIMS OF CDTMS FACULTY HOLLAND PARK SCHOOL (1984)

1. The pupils' physical, emotional, moral and social development.
2. The development of pupils' awareness and the awareness of the relevance of areas within the faculty to society and the historical context of these areas.

3. Encouraging the pupil to ask questions and search for her or his own answers.
4. Encouraging and providing conditions through which a pupil may develop and continue to develop.
5. Helping pupils become interdependent and independent in their learning.
6. Encouraging and supporting individual teachers to enable them to work towards the achievement of these aims.
7. To develop anti-sexist and anti-racist attitudes by both staff and students.

SCIENCE

For the purpose of this article we focus on the science area, although this is in a sense a false separation, as we are working towards making science part of an integrated curriculum which crosses traditional subject boundaries.

'The Science Department' was initially suspicious of a common curriculum. We feared that if teachers were able to teach the 'same sort of science to all', larger classes would be the result. At least the option system meant several different subjects and many smaller classes. However, commitment to the idea of an integrated curriculum enabled us to overcome our misgivings. The removal of subject overlaps would allow time to explore with students some of the fundamental concerns as to what constitutes science, and who it is for. We considered this to be important; if students were to study the subject, they ought to be able to think critically about it.

We also had a lowered pupil/teacher ratio after the amalgamation because of strong union action by the staff of the school. This became important in the science context, as it allowed the use of 'support' staff when science was being taught.

An early meeting with the Secondary Science Curriculum Review highlighted the difficulties we would face. We agreed (or thought we had agreed) that our new course could not be content-led. We needed more than just a list of topics to be covered. We wanted classroom strategies that would allow for a greater degree of control by students over their own learning. We felt that small-group work would ease the path of co-operative learning, and that anti-racist and anti-sexist work must be an integral part of the course. None of the staff had been trained for anything other than 'chalk and talk'. And as for anti-racist and anti-sexist education... where do you start? (What *is* it, anyway?)

ANTI-RACIST EDUCATION

In 1983 the ILEA's anti-racist policy was introduced and soon raised questions about what it meant. Anti-racist education could not be reduced to 'equal opportunities'. Teachers who for years had been committed to promoting equal opportunities began to take stock of the effect of racism on the life-chances of the children in their care. This led them to begin to examine the institutionalized nature of racism in British society, and to analyse its roots in the history of Britain's relationship with its former colonies.

Even before the introduction of ILEA's anti-racist policy, our school had an anti-racist working party, made up of black teachers and white anti-racist members of staff. This meant that the school had a history of discussing the issue, and a small group of people had begun to develop appropriate ideas and expertise. Some members of the CDTMS faculty had been involved in this work.

By that time the curriculum had not been fully analysed in terms of its role as a vehicle for racist ideology. This analysis is still only in its early stages. However, this process led the science staff towards a critical appraisal of its work. We began to devise strategies for exposing racist ideology wherever we were aware of it, and for teaching science in a way that did not place black/female/working-class students at a disadvantage.

Three members of the science staff began to meet anti-racist teachers from other schools, on Saturday mornings, in a group which eventually became 'The Association for Curriculum Development in Science'. The group had been drawn together by two members of the ILEA anti-racist strategies team –

Science Curriculum Innovation 155

seconded workers who had been brought in to work out 'anti-racist strategies' for secondary education.

This group provided support for its members. We shared ideas and materials. We wrote seminar papers for discussion. Anti-racist education was our focus, but this raised many other issues which gave us an insight into the philosophy and history of science. A growing understanding of science as a social process increasingly improved our work within the school.

PEDAGOGY

Although we had begun to experiment with group work and collaborative learning during the 1970s, we had not yet evolved a coherent philosophy of non-oppressive pedagogy. We are still working towards this. Some members of the science staff are convinced of the need for working in ways which challenge traditional power relationships through classroom organization as well as curriculum content. Others feel insecure and out of control if their lessons do not mirror the traditional school classroom of their own youth.

There is need for much more discussion of ideas on the aims of education and politics of schooling, and for sharing ideas on the practical implications of equal opportunities and anti-racist objectives. The philosophical underpinnings and practical techniques are neither fully understood nor fully subscribed to by all members or the science staff. It is largely for this reason and others that our early experiments with innovative curriculum design were a failure – not a total failure, it must be stressed, but certainly not successful. Failure or not by our own declared standards (see our 1984 Aims), the exam results were no worse than in previous years. We felt that we had not failed the pupils any more in trying for new methods than we had been failing them in the past, with the old.

THE 'ISSUES-BASED' SCIENCE COURSE

Now we describe the science course that was introduced in 1984, and explain how it was supposed to work: we also try to explain why it did not.

Initially we undertook to look at the course as a series of units stretching from the third to the fifth year. This in itself was innovative, because previously the lower-school course had been thought of as separate from the examination course followed by fourth- and fifth-formers. In the new course the third-year and fourth-year work was co-ordinated and organized by two different people. This led to problems, which will be outlined below.

Our discussions led us to consider an idea from Jan Harding, who was working with postgraduate students at Chelsea College. They had prepared a rationale which we thought had promise; in particular we thought that posing questions on 'Issues' had merit. 'How do you solve the World Food Problem?' Science and Technology may throw some light on the question, but essentially the answers lie in social, economic and political action. One advantage of this 'Issues-based' approach is that no teacher is able to pose as an 'expert'. This had implications for the power relations in the classroom in reducing the power gap between 'teacher' and 'taught'.

One major advantage of the ten years' previous discussion on equal opportunities had been that the science staff were united in their agreement on the damaging effects of examinations on the majority of students and on education. We were determined that where possible we would not have our curriculum content led by the nose, but we could not ignore the examination system altogether.

In asking 'equal opportunities for what?', we discovered that much of the rhetoric was to do with getting all children to compete equally for the small number of jobs as, say, nuclear physicists. We had begun to be very critical of the notion that it is the purpose of a comprehensive school to give 'equal opportunities' if all this means is that everyone should have the same chance of being one of the few people who end up with A levels. This kind of 'equal opportunities' means that the school is structured to serve the selection process for higher levels of education. Curriculum content is similarly determined.

In rejecting that narrow definition of equal opportunities, we were not in a position also to reject the examination system. What we had to do, we felt, was to provide a useful and relevant education for all, while also giving the chance of O and A level exam success. We studied existing syllabuses in Physics, Chemistry, Biology and the proposed Joint syllabuses. By using

158 Anti-Racist Science Teaching

SOCIAL, ECONOMIC, ENVIRONMENTAL AND CULTURAL IMPLICATIONS OF CHEMISTRY

Finite nature of resources
Pollution
Food supply
Chemotherapy
Misuse of substances
Radioactivity
Chemists as people, etc.

MATERIALS:
their behaviour,
composition sources,
uses

ACTIVITIES
Designing experiments
Testing hypotheses
Measuring, observing
Recording, communicating
Collaborating

IDEAS – MODELS – PATTERNS – THEORIES
Patterns – solid, liquid, gas, pure substances, mixtures, elements, compounds, metallic and non-metallic characteristics, reactivity series, periodic table, acids, bases, salts; pH

Atomic structure and bonding; radioactivity

Energetics
Reversible reactions
Speeds of reactions
Oxidation and reduction
Electrochemistry
Quantitative nature of chemical reactions

Chemistry from 'Issues' at 16+   — *Core*

some of Jan Harding's Networks as a base, we thought we could expand the network idea to cover all the syllabuses and so ensure that examination work would be covered.

Energy Matters
Here we present modules from the issue-based course. 'Energy matters', like the other modules, could be used as the basis for an examination of the social context within which science is done, as well as the means of enabling us to explore ideologies of racism, sexism and social class in relation to science.

With the miners' strike still taking place, it seemed appropriate when discussing energy to bring into classroom discussion the role of oil companies, the nuclear industry and its published materials for schools, and various alternative forms of energy. This led to an examination of the relevance of energy policies to global conservation issues.

Some teachers tried to connect up the situation of South Africa and its domination of Namibia with the miners' struggle. Others examined social and political aspects of nuclear energy, and related Britain's nuclear policy to the miners' strike. However, at this time teachers' industrial action prevented departmental meetings, so it was left very much to individual teachers to work out what to do, with little discussion between colleagues.

RUNNING THE NEW COURSE: PITFALLS AND PROBLEMS

The new course worked differently for third- and fourth-year groups. It is instructive to try to analyse the reasons for the differences.

Control of a third-year class is comparatively easy, because groups are small. Now that science is compulsory in the common-core curriculum, we have to use the school's smaller-sized laboratories to accommodate everyone. It happened that third-year classes, accommodated in small labs, had to be taught in groups of fifteen. Most of the classes were taught by their form tutors, or staff who knew them well, having taught them in previous years.

THE RATIONALE

On the rationale for 16+ 'Chemistry from Issues' Course, Jan Harding writes:

☐ Chemistry is studied in the fourth and fifth year by significantly fewer pupils than is physics or biology. Chemistry had little or no place in the curriculum of the Secondary Modern school. This indicates that there is no strong conviction about the place of chemistry (as presently contained in examination courses) in the education of any but those who may wish to use it in a formal way for qualification or employment.

☐ The essential nature of chemistry has been presented as the body of concepts and principles which account for the facts of chemistry. Curriculum development in the 1960s has strengthened this emphasis. But much of the formalized structure of chemistry may be beyond the reach of most pupils in the eleven-to-sixteen part of the school (Shayer, 1978), especially if presented within a framework dictated by conceptual structure. Moreover, this view of the chemistry may be a misrepresentation of chemistry as it exists and is practised.

☐ Chemistry is supremely a technological activity: we use, modify, purify and create materials and are concerned with the discipline of chemistry as a tool for enabling these activities. This view of chemistry finds support in the way many major 'advances' in chemistry occurred, and is just as valid a view as one which emphasizes a 'knowledge for its own sake' view.

☐ Evidence from personality and maturation studies (Head, 1980) suggests that to attract able, thinking and reflective adolescents to the physical sciences, their presentation in schools should be changed to relate them more closely to the issues that young people see as important in their own and others' lives. Girls, especially, are more committed to science and technology if these connections are stressed.

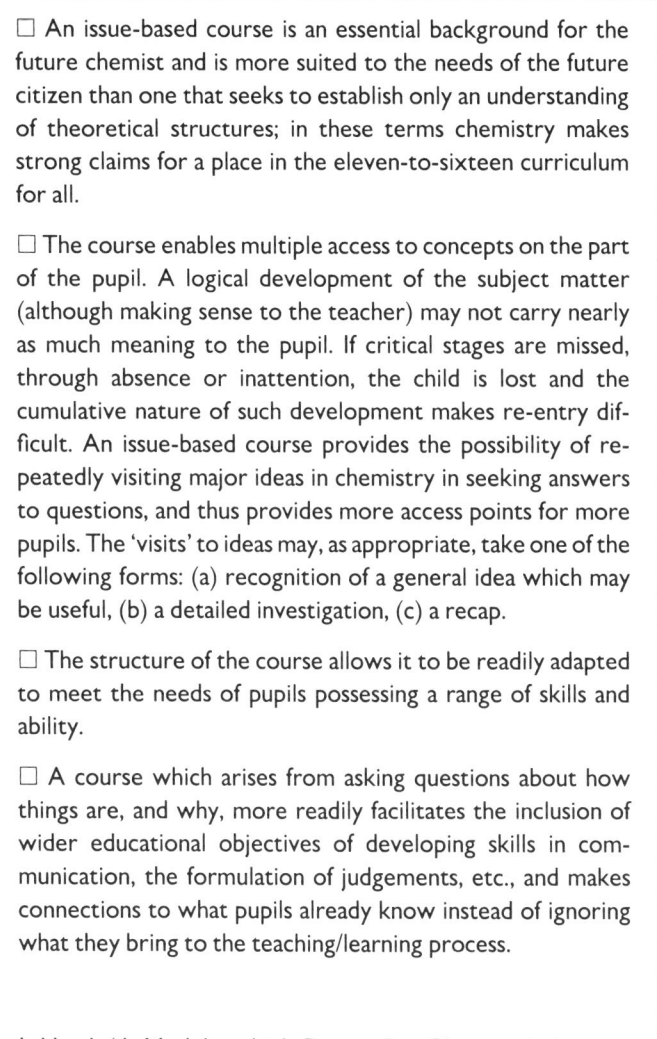

☐ An issue-based course is an essential background for the future chemist and is more suited to the needs of the future citizen than one that seeks to establish only an understanding of theoretical structures; in these terms chemistry makes strong claims for a place in the eleven-to-sixteen curriculum for all.

☐ The course enables multiple access to concepts on the part of the pupil. A logical development of the subject matter (although making sense to the teacher) may not carry nearly as much meaning to the pupil. If critical stages are missed, through absence or inattention, the child is lost and the cumulative nature of such development makes re-entry difficult. An issue-based course provides the possibility of repeatedly visiting major ideas in chemistry in seeking answers to questions, and thus provides more access points for more pupils. The 'visits' to ideas may, as appropriate, take one of the following forms: (a) recognition of a general idea which may be useful, (b) a detailed investigation, (c) a recap.

☐ The structure of the course allows it to be readily adapted to meet the needs of pupils possessing a range of skills and ability.

☐ A course which arises from asking questions about how things are, and why, more readily facilitates the inclusion of wider educational objectives of developing skills in communication, the formulation of judgements, etc., and makes connections to what pupils already know instead of ignoring what they bring to the teaching/learning process.

J. Head, 'A Model to Link Personality Characteristics to a Preference for Science', *European Journal of Science Education*, vol. 2 (1980), 295–300.
M. Shayer, 'The Analysis of Science Curriculum for Piagetian Level of Demand', *Studies in Science Education* 5 (1978), 115–30, Leeds.

16+ (FOURTH AND FIFTH YE[AR])

DRAFT

Entry Issues	Questions that can be asked
ENERGY MATTERS	
A. 'Save it!' stickers Nuclear energy: no thanks! Poisoned by fire!	Why are we 'short' of energy? What is 'nuclear' energy? What is radioactivity? Which fuel is best? What happens in combustion? Is it OK to use up all the coal and oil? What else is oil used for?
HUNT DOWN THE LEAD	
B. Lead in petrol. Lead in water. Lead in dust.	What is the evidence for toxicity? Why is/was it used in the first place? Can we do without it? What is the cost? How else does lead get into us? Why are some waters safe? Why does lead behave as it does? What are the alternatives?
WATER: WHAT'S IN IT?	
C. Acid Rain. Smog. Toxic effluent.	What does water normally contain? Where does the acid come from? What does acid do to the soil? Why is the sea salty? What are salts?
THE CARBON DIOXIDE STORY	
D. Destruction of ozone layer. 'Greenhouse' effect.	What's in the atmosphere? What effect does it have? Has it always been the same? How did the oxygen get there? How is it maintained? What happens to CO_2?
FOOD & FERTILIZERS	
E. Beware nitrates! World food problem. Butter mountains.	What do foods contain? How do they get there? What do foods need to grow? What grows where? Is there a shortage? What is fertilizer? Which is the best? How are they made? How much do they cost? Are there side effects? What's in a diet?

The development for this proposal was carried out by the PGCE Chemistry students (1982/3) at Chelsea College, Centre for Science and Mathematics Education, in

CHEMISTRY COURSE FROM 'ISSUES'

Chemistry content that could be included

Atomic structure: energy in nucleus – atoms change. Combustion: energy obtained from rearrangement of atoms which are unchanged. Concept of molecule. Measurement of heat of combustion of carbon, methane, etc. Products of combustion, incomplete combustion of plastics. C-Chemistry: hydrocarbons plastics, fermentation (alcohol). Fuel cells, other cells.

Internal combustion engine, 'anti-knock', combustion hydrocarbons. Properties of metallic lead, ease of extraction, use by plumbers (*in situ*). Density (relative atomic mass). Hardness, pH of water, Copper, plastics.

Acidity/alkalinity, pH, pH salt solutions, water cycle, natural waters, hardness of water, composition of sea water, 'salts from the sea', halogens (elements, compounds, ionic bonds, Group 7), $S + 2O_2 +$ sulphuric acid, nitric acid. Preparation of salts (ionic equation) simple geology, composition of soils.

Photosynthesis, respiration, combustion, carbon cycle. Stability of CO_2, heat of formation CO_2 is water, action on limestone (equilibria, reversible reactions), C^{14} dating (isotopes).

Composition of foods, N cycle, sewage treatment, Haber process (rates, equilibria, catalyst), soils, pH, chemistry of ammonia, nitric acid. Food and medicine.

---

collaboration with co-tutors in teaching practice schools and their Centre supervising tutors Drs Erica Glynn, John Head, Jan Harding.

The day-to-day content of the science course was decided to a large extent by the teacher of the class, although all work lay within the framework of a clearly articulated syllabus. It happened that the team of people teaching this particular year group favoured a structured 'worksheet' approach to their lessons. We had previously been following a 'Science at Work' syllabus with the third year. Some teachers were not happy at losing the flexibility that this course provided, where plenty of 'extras' could be inserted; others were happier that a more rigorous science scheme was now employed.

Although our discussions after amalgamation applied to all year groups, greater effects were seen with the fourth-year group, and inevitably comparisons were made between that year group and the third. Our 'issues-based' fourth- and fifth-year course seemed to compare unfavourably with the course on offer to the third year.

Resources
The subject options were timetabled in such a way that the ten fourth-year classes were split into two halves: the CTDMS faculty dealt with one half at a time. Classes had four lessons per week, each an hour long.

Five tutor sets and five teachers were timetabled as teacher-tutors responsible for registering, record-keeping and science teaching. Extra staff were allocated to each half year as support staff, in order to improve the student–teacher ratio. Generally speaking, 'support' did not have the desired effect; support staff did not take on much responsibility, probably because, with one exception, these staff were not timetabled for all four of the lessons. (This was something we put right the following year.)

During the summer before the course started, we ripped out a lecture theatre room and partially converted it into a resources area. All our old materials, texts, worksheets, videos, etc., were housed here. The intention was that students would leave their laboratories to investigate or collect their own resources. Generally this worked quite well (apart from the quality of the resources), but such work does put a strain on resources. For example, until we know that a certain text 'works' in a given situation, it cannot be brought in as a standard text, even if we are able to afford it.

Industrial Action
The very nature of our aims in doing a new course meant that as teachers we needed to talk amongst ourselves in depth. This we did to a certain extent in formulating the aims. We knew that we wanted to bring in far-reaching changes of our teaching methods, and that this required massive changes by teachers themselves. Then the industrial action started and we could have no more formal meetings.

We had already discussed concepts which were new for us – small-group work, projects, less didactic 'chalk and talk', a much more open approach to the very nature of science – but we really did need a continuous dialogue amongst ourselves. When meetings were eventually reinstated, some staff did not attend, others did not prepare materials and we were left in the position yet again of preparing a sort of 'teacher-proof' packet of materials – in our experience, the very thing almost guaranteed not to work.

Teacher Insecurity: Student Insecurity
After we started our 'team approach', some members of the teaching team felt worried and threatened almost immediately by the feeling that they were not in full control of what happened in the class. The idea of pupils taking greater responsibility for their own learning is difficult for some teachers to handle. It must also be added that the vast majority of the teachers with the fourth-year group had not taught these classes before.

Some interesting developments occurred. Some 'moan sessions' took place. These were very nearly full of destructive criticism by the staff present – often those who had not attended team meetings or work-preparation sessions.

Students had become responsible for organizing their own experiments: if they needed materials they had to order equipment, videos, slides, etc., from the technician team, whose structure, incidentally, had also been completely changed by the amalgamation. The amount of practical work by both boys and girls decreased dramatically.

We are not quite sure why practical work faded. Was it really too much of an effort for students even to order a video or experiment for the next lesson? For students who would leave school in eighteen months, was it really too difficult to pre-plan the work even a day or two ahead? Were the experiments we wanted done just the same old experiments 'proving' things that students had just read about? Did they reject practical work because it failed to challenge them with new ideas?

Many of the students adopted a worksheet syndrome, forever demanding more sheets with easily answered questions. They never really accepted the idea that useful work could be done through discussion. A few demanded the security of textbooks and copied or paraphrased whole sections. After two terms nearly all the stock of texts (mostly inappropriate ones, anyway) had vanished from the resources area and been taken home. Was this because the students believed textbooks to be the sources of 'real' science?

Most teachers slowly but surely reverted to traditional methods of teaching and classroom organization. A few openly showed dislike of the course and sabotaged it; some adopted a 'I will give a lecture – take it or leave it' approach. Here, as we had an 'open-class' policy, some students did indeed leave it to join other laboratories; some of the 'exam-orientated' fourth-year students joined up with the lecture approach. Some teachers also suffered from the worksheet syndrome and gave out reams of worksheets.

In its first year of operation, the 'issue-based' approach to science clearly had not worked well for either teachers or pupils.

A REAPPRAISAL

After the end of industrial action in 1984, staff meetings could resume. We moved more into faculty development, meeting not as science teachers but as teaching teams for maths, design and technology, food, fabric and science, looking at the eleven-to-sixteen curriculum. Many of us realized that anti-racist and anti-sexist initiatives must take seriously the issues of practice and control in classrooms; and other parts of the faculty are also becoming more aware of this.

Towards the end of the initial year of the new course, it was decided that the third-year section needed 'polishing' and that the fourth-year students should choose two areas of specialization in science for their fifth year. This, the feeling went, would sort out the problems and give staff some breathing space. The areas of 'specialist work' would be those for which students would take the major responsibility, working in the 'issues-based' way, using the resources bank and planning their own programme of learning. For the rest of the time they would be taught more formally.

This solution would, we hoped, give all teachers a chance to teach in the way they felt most confident – whether this is in the traditional 'talk and chalk' mould, or in the new mould of issues-based work. Setting up a framework where traditional formal teaching is paralleled with more progressive methods has advantages: it gives teachers time to develop new skills and expertise without losing the security of their old skills and expertise. It allows pupils to gain confidence in new styles of learning without losing the security of the old.

Some groups of students have taken to the idea of having more 'space' for themselves. It remains to be seen how these pupils adapt to a formal teaching situation again. Early indications are that those teachers who had thought of trying a less formal approach with a formal curriculum are not doing so. The major reason they give is the short time available to finish the work on the syllabus.

When we reached the last half-term of the 1984/85 academic year, no formal meetings took place outside school hours, but we did have the extra time that the loss of exam classes gave us. This made it possible to evaluate the first year of the course, and make plans for its second year. This process was slowed down once again as a result of the conflict between champions of the 'traditional approach' and what has come to be known as the 'issue-based approach'.

The majority of the staff involved in the issue-based scheme had been involved in the discussions on the aims of the faculty; in that context it was relatively easy for those in favour of the new approach to 'win' the argument. The main reason for this is the fact that everyone now accepts that science is not value-free. However, many concessions had to be made to traditional teaching. The alternative scheme had to be 'watered down' to allow reluctant teachers to gain confidence. Many of the ideas now accepted by the majority would not have been accepted had the fourth-year experience not been implemented during 1984/85.

EVALUATION OF THE PILOT SCHEME

First, from the pupil's point of view, the problems seemed to be as follows:

1. The issues network with which they were presented was too open-ended and too comprehensive for them to find security, comfort or understanding in it.
2. The set of issues from which they were required to choose were not necessarily 'their' issues. These issues often appeared to them to be far too 'national' or 'international', removed from their own experience. The students often felt that they could not hold views on important matters such as water control, disease, or malnutrition.
3. The organization of the scheme appeared to be too structured and too individualized for interaction to take place frequently and meaningfully in small groups.
4. Pupils who lacked organizational skills often lost out, as the teachers were stretched by the more confident students, who demanded more and more work.
5. Quite often resources and materials used were inappropriate. Pupils unable to find suitable materials reverted to what they had been trained to do so well in the first three years – answering questions on worksheets.
6. The majority of pupils have been trained to feel that the

Science Curriculum Innovation

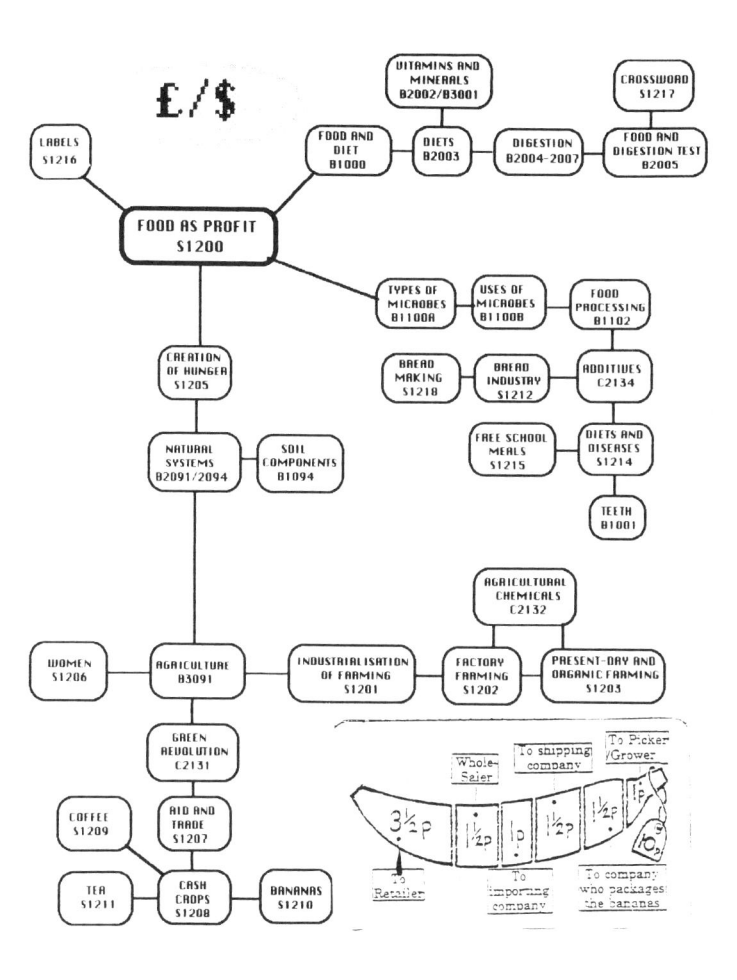

Holland Park School's new course material has become a focus of official scrutiny at ILEA and DES level. In 1986 ILEA science inspectors carried off copies of all the sheets from the syllabuses, like this 'issues network'.

amount of written work in their books is proportional to the amount of learning that has taken place.

7. Students were sometimes unable to answer the questions posed because their previous learning experience had led them, falsely, to expect that all questions have one right answer.

The lack of commitment from teachers could be attributed to the following:

1. If the pupils left a class to visit the resource centre or another lesson, the teachers felt loss of control over a 'class'. After all, they are trained as 'classroom teachers'. The sense of insecurity and the loss of power can be very threatening.

2. The network looked too vast to manage, especially as it contained materials on subjects they did not specialize in. It proved difficult for some teachers to say to a pupil, 'I don't know'.

3. Teachers were sometimes unable to provide suitable resources and may have been ignorant of resources available. Some teachers did not want the pupils to challenge the work set or the values presented by that work.

4. Teachers were frightened that the syllabus for an exam was not being covered; easy methods of monitoring pupil activities and progress were not available.

5. The classes often had all the individuals working alone; this meant that twenty-five or more different activities would have to be dealt with in each class!

CHANGES AFTER EVALUATION

In order to overcome as many of the above problems as possible, the following decisions were taken:

1. A compromise syllabus was adopted. Initially, this is to be a GCSE in Science – Biology, Chemistry, Physics – for double certification (two exam certificates). Most syllabuses on offer in Integrated Science are similar in content. If we decide to change the syllabus in the future, this will not pose problems. Unfortunately, this compromise also allows the flexibility to return to subject-specific exams.

2. A contents network was drawn up, similar in style to the

Science Curriculum Innovation

Maths SMILE (ILEA's Secondary Mathematics Individualized Learning Experiment). The syllabus was divided up into discrete topics of work and arranged in three levels of difficulty. This would allow the teachers to monitor the areas of syllabus covered by individual pupils. Insecure pupils would like this too!

3. The Issues networks would be curtailed substantially, and the number of paths through a network limited. Networks would be amended and modified to take account of pupil interests and confidence.

4. All aspects of the network would be suitably resourced; pupils would have to follow a complete path on the network. At the end of this there would be a variety of 'assessment' tasks.

5. Pupils would work in small groups rather than individually. This would allow for greater interaction amongst the members of the group and make it easier to adopt strategies that would allow greater interaction between the groups.

In 1985/86 we entered another year of industrial action. Once again all meetings were halted. We spent time formalizing (what was meant to be) a less closed approach to science. Different parts of the school had differing perceptions of what occurred in science. Staff reactions ranged from almost pleasure at seeing a failure to almost pleasure at seeing the attempt, the reaction depending largely on where they got their information on what was taking place. The fourth-year students were being taught by people who knew them relatively well, and the team approach was curtailed. New ideas needed to be discussed and used by all teachers so that experiences could be shared and analysed, but all that had to remain at the individual level until the end of the industrial action.

LESSONS LEARNED

Our description might suggest that the project had been a massive failure, but we felt otherwise. We learned a lot. Perhaps the fourth-year group was the wrong place to start issue-based learning. However, with our first-year students we started another problem-solving open-ended science scheme linked to Design and Technology. This also had limited success, depending on the teachers' willingness and confidence to abandon the security that textbooks or work cards provide. Generally, the faculty is not yet prepared to abandon it.

With the growth of teacher confidence and the development of new styles of teaching and learning, we are convinced that the new scheme will be successful. As pupils get older, their experience of individualized learning and group work will stand them in good stead. By the time the present first year reaches the fourth, their educational experience will be very different from that of the original 'chemistry from issues' group. They are less likely to be thrown into fits of insecurity by the fact that they are expected to take responsibility for organizing their own day-to-day work in negotiation with a teacher. They are less likely to have expectations of being told what to do in minute detail. Already this year group is confident and capable; pupils do not constantly look to the teacher for a 'right answer'. They are prepared to research 'answers' for themselves; they accept that, for many complex questions, there is no one single 'right' answer – though some solutions to a problem may be better than others.

Although some of our methods did not work well in the 'Science from Issues' course, the science staff feel that our aims are sound. We do not intend to give up – merely to move more slowly, to take one step back in order to be able to take two steps forward in the future towards an anti-racist science. Most science staff would now agree with official reports on maths and science teaching (e.g., the Cockcroft Report), which stress the limitations of conventional subject boundaries in failing to give students a critical understanding of the subjects and their social role; clearly conventional subject boundaries hinder anti-racist and anti-sexist education.

If ILEA policies on racism, sexism and social class are to be put into practice in all subjects, it is necessary for teachers and students to explore the links between subjects and to examine critically curriculum content and its social origin. There is an urgent need to develop curriculum content and pedagogies

Science Curriculum Innovation

which expose and challenge the ideologies of racism, sexism and social class. It is equally important that learning materials reflect and value the countries and cultures from which Britain's black populations originally came.

An anti-racist perspective in science must address the following issues:
- the colonial history which underlies racism in British society;
- the devaluation of 'non-Western' cultures;
- development and underdevelopment;
- racism in contemporary society.

The development of learning materials needs to be informed by consideration of the following issues:
- the links between curriculum content and its social context;
- the need to examine the operation of the school as an institution, and in particular the significance of pedagogic styles on interrelationships between people and on pupil performance;
- the language of textbooks (and other learning materials), which often makes subject content inaccessible to students and reinforces existing ideologies of racism, sexism and social class;
- the need for learning materials and classroom strategies which encourage collaborative and individualized learning and small-group work; and
- the educational advantages of cross-curricular links and subject integration.

In Holland Park School we have only just begun what is an enormous task of curriculum development. It is clear that our path will be fraught with difficulties, but we are determined to overcome them. In turning 'equal opportunities' rhetoric into reality, we are caught up somewhat in the past – which slows our progress. But we are sure of moving forward, secure in the knowledge that whatever we do, at least we are posing and attempting to answer honestly the questions raised in discussion about equal opportunities. We are clear that we should not be simply helping students to compete more fairly for the few available positions as scientists. And we are exploring how students can overcome deference to technical experts –

including teachers – through group work and independent investigation.

As we redefine equal opportunities in that way, we can also raise the question of equality. Given that racism is rooted in structured inequalities of power, no teaching method in itself can change those wider power relations. But it can teach students how to recognize and challenge them, especially when disguised as technical expertise. In that sense our approach can motivate students to learn skills they can use to seek political equality.

(Written spring 1986)

Postscript

As we enter the third consecutive year of industrial action, we face bleak prospects: no report writing, no parent meetings, no cover for absent teachers, no after-school voluntary activities, no in-service training, no pay settlement. Meanwhile our school is dealing with an asbestos removal problem at the same time as introducing new GCSE syllabi. The continuing impasse between the teachers' unions and the government has resulted in minimal teacher participation in formulating the syllabi, much less in raising anti-racist issues around them. Perhaps the worst result has been the lack of opportunity for teachers to discuss the situation in our school. It is difficult for us to give mutual support in developing new teaching methods without that costing time. By default, some colleagues have found ways to preserve their health and sanity: they return to presenting traditional class lessons, or they organize their class so that all the students do the same work at the same time. Nevertheless, the base of our new course remains in place; despite the limited time available, we are producing new resources to add to it.

(Spring 1987)

Race

INTRODUCTION

The racist content of modern Western science has a long history. The rise of British colonialism was accompanied by scientific approaches which sought a natural legitimation for British domination of African, Caribbean and Asian people. Of course, that colonial-era scientific racism was so blatant that little of it survives in its original form in science today. Nevertheless it is worth examining that legacy, for at least two reasons:
☐ Seeing such a blatant form of scientific racism, and its close connection with both class and race exploitation, can help us to see today's more subtle forms.
☐ Seeing how racism constituted the world-view of several scientific disciplines can help us see that the racism was no mere distortion or abuse of science, but defined what counted as science.

For those reasons we include Peter Fryer's article on pseudo-scientific racism. At the same time, we would hesitate to label his revealing examples as mere 'pseudo-scientific racism', as this *was* the science of that period. Then, as now, there is no science other than the science that gets done.

Once having seen the historically racist foundation of British science, the reader would do well to connect that history to the present. Nineteenth-century racism has left a legacy deeply embedded in British culture today: in literature, in the fields of history, psychology, biology, anthropology and geography, as well as in the courses for teacher training. It finds more blatant expression in newspaper and television reports, in television chat shows and in books and comics written for children. In Britain's schools, these racist ghosts of the past have tended to cripple the living, for the present and the future. They will continue to do so unless teachers expose and challenge that legacy.

A crucial element of present-day racism is the supposedly biological basis for subdividing the human species into separate 'races'. Emphasizing obvious differences in certain physical traits, the entire language of 'racial differences' assumes that innate mental traits correspond to those physical traits. Here Richard Lewontin shows that the racial categories have themselves changed historically in arbitrary ways. He also shows that the traditional racist model has been superseded by even mainstream science, which recognizes far more genetic variation among individuals within each 'race' than between the 'races'. Since human 'racial' differences are only skin-deep, the use of racial categories must be based on social preconceptions which we project on to physical traits. Of course, this will continue to happen, regardless of scientific facts to the contrary, as long as science continues to operate in a racist society.

PSEUDO-SCIENTIFIC RACISM

PETER FRYER

By the 1770s racism was firmly established in Britain as 'a principal handmaiden to the slave trade and slavery'. The British slave trade was ended in 1807; slavery in 1833. Could racism now be dispensed with? By no means. It was too valuable. A new basis and new purpose for it had emerged. It was to become a principal handmaiden to empire. The culminating stage in the rise of English racism was the development of a strident pseudo-scientific mythology of race that would become the most important ingredient in British imperial theory.

This mythology arose in the 1770s, precisely when the British government first had to face the problem of ruling a territory with 'natives' in it. In 1773 the Regulating Act asserted parliamentary control over the East India Company for the first time; Warren Hastings was appointed first governor-general of Bengal and a supreme court was set up in Calcutta. In the following year Edward Long's *History of Jamaica*, the pivotal book in the turn to pseudo-scientific racism, was published. Pure coincidence, of course. But the timing turned out to be crucial for the subsequent history of the British Empire. From the 1770s onwards the empire and the pseudo-scientific racism that served it developed side by side. Even the cosmetic version of the doctrine – the idea of 'imperial trusteeship' for the betterment of 'backward peoples' – took shape in the debates over the abolition of the slave trade in the years before 1807. From the 1840s to the 1940s, Britain's 'native policy' was dominated by racism. The golden age of the British Empire was the golden age of British racism too.

Long's book, furnishing ready-made arguments for the belief that black people were innately inferior to white people, was widely read, and widely accepted, by the scientists of his own time and for some forty years after his death in 1813. These arguments crop up again and again in the writings of later polygenists of high scientific repute. But Long was not the sole source of pseudo-scientific racism. It was nourished also by a series of theories and 'discoveries' in the biological sciences of the eighteenth century: notably, the infant science of anthropology. This was the great century of classification, and it was the Swedish botanist Carl Linné, generally known as Linnaeus, who laid the basis for the modern classification of plants and animals. He was first to call us *Homo sapiens*, and he arranged us in a hierarchy largely based on skin colour, with whites at the top. In the 1758 edition of his *Systema naturae* he gave thumbnail sketches of, amongst others, European Man and African Man. Here is how these descriptions were translated into English in 1792:

> *H. Europaei.* Of fair complexion, sanguine temperament, and brawny form . . . Of gentle manners, acute in judgement, of quick invention, and governed by fixed laws . . .
> *H. Afri.* Of black complexion, phlegmatic temperament, and relaxed fibre . . . Of crafty, indolent, and careless disposition, and are governed in their actions by caprice. – Anoint the skin with grease.

It was a pupil of Linnaeus who, relating an experiment by the French naturalist Réaumur, in which a rabbit allegedly fertilized a hen to produce chicks covered with fine hair instead of feathers, commented that 'the most frightful conclusion could be drawn from this; . . . one would have reason to think that the Moors had a rather strange origin'. Clearly, all those travellers' tales had become 'thoroughly cemented into Western thought'. The Swiss naturalist Charles Bonnet wrote in 1764 of the 'prodigious number of continued links' between 'the most perfect man' and the ape, and left no doubt in his readers' minds as to the identity of 'the most perfect man': 'Let the flat-faced African, with his black complexion and woolly hair, give place to the European, whose regular features are set off by the whiteness of his complexion and beauty of his head of hair. To the filthiness [*malpropreté*] of a Hottentot, oppose the neatness of a Dutchman.' This brand of 'scientific' thought was reflected in the writings of the British MP and Board of Trade official Soame Jenyns, whose treatise 'On the Chain of Universal Being' also put 'the brutal Hottentot' at the bottom of the scale but preferred Newton at 'the summit' to any Dutchman, however neat.

CRANIOLOGY

A major but not always deliberate contribution to pseudo-scientific racism was made in the eighteenth century by pioneering students of the human skull. The 'father of craniology' was a German professor of medicine called Johann Friedrich Blumenbach. Co-founder with Buffon of the science of physical anthropology, Blumenbach disliked the 'Chain of Being', divided humanity first into four and then into five varieties, denied that any of these varieties was inferior to others, and denied that Africans could not acquire learning. A friend of Ignatius Sancho by correspondence, he collected a library of books by black authors. But he also collected human skulls from all over the world, and it was one of these skulls, from the Caucasus in Russia, that led him to suppose that Europeans came from that region, to coin the word 'Caucasian' to describe the white variety of humans, and to prefer this 'most beautiful form of the skull' to the two extremes furthest from it, skulls which he called 'Mongolian' and 'Ethiopian'. Blumenbach's method of studying skulls was to stick them between his feet and examine them from above to see what shape they were.

Blumenbach's method did not satisfy the Dutch comparative anatomist Pieter Camper, whose 'facial angle' – an expression he himself does not seem to have used – was, in a sense, the springboard of modern craniology. Camper's 'facial angle'

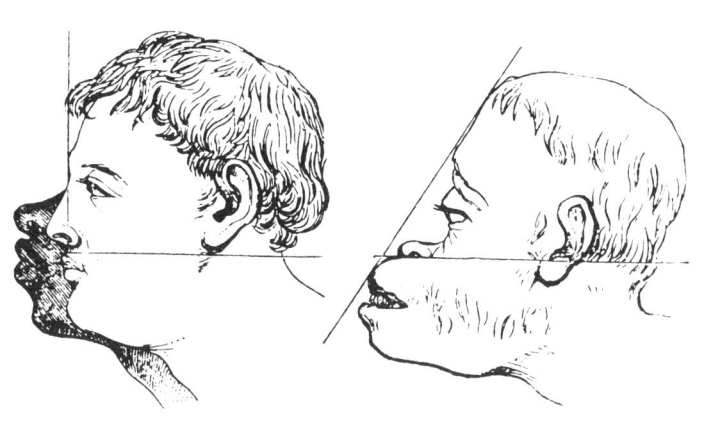

[*Profile of Negro, European, and Oran Outan.*]

Examples of nineteenth-century science

measured the extent to which the jaw juts out from the rest of the skull. A wide angle was thought to indicate a higher forehead, a bigger brain, more intelligence, and a more beautiful appearance. According to Camper, the angle grew wider as one went from Africans, through Indians, to Europeans. And it was 'amusing', he wrote, to contemplate an arrangement of skulls on a shelf in his cabinet, placed 'in a regular succession' from apes, through Africans, to Europeans. Camper found 'a striking resemblance between the race of Monkies and of Blacks'. (As W. D. Jordan points out, if amount of hair had been chosen as the criterion of ranking, Africans would have come out on top, Indians in the middle, and Europeans at the bottom, next to apes: 'When Europeans set about to rank the varieties of men, their decision that the Negro was at the bottom and the white man at the top was not dictated ... by the facts of human biology.') The German anatomist Thomas Soemmerring, pupil and friend of Camper, Blumenbach, and Goethe, also tried to prove that Africans' skulls were intermediate between those of Europeans and those of monkeys, and that the African's brain was 'smaller than that of the European'.

Meanwhile the Scottish lawyer, polymath, and 'commonsense' polygenist Lord Kames was pioneering a racist interpretation of society by claiming that ever since the Tower of Babel humanity had been divided into different species, each adapted to a different climate. 'The black colour of negroes, thick lips, flat nose, crisped woolly hair, and rank smell, distinguish them from every other race of men', he wrote – and by 'race', as he made clear, he meant 'species'.

The scene was now set for the influential Manchester physician Charles White to lecture to the Literary and Philosophical Society of that city in 1795, quoting United States President Thomas Jefferson's 'suspicion' that Blacks were inferior to Whites, and concluding with a declaration of white superiority in terms of beauty, posture and intellect.

SCIENTIFIC THOUGHT

With Charles White's lecture, published in 1799 as *Account of the regular gradation in man*, English racism may be said to have come of age. Its later development, throughout the nineteenth century, presents a picture of remarkable diversity and complexity, which there is room here only to summarize. Three facts will help to guide us through this maze of theories.

First, racism was not confined to a handful of cranks. Virtually every scientist and intellectual in nineteenth-century Britain took it for granted that only people with white skin were capable of thinking and governing. Even the distinguished ethnologist James Cowles Prichard, humanitarian and monogenist though he was, insisted on a relation between the 'physical character' of West Africans and their 'moral condition'. The Igbos, 'in the greatest degree remarkable for deformed countenances, projecting jaws, flat foreheads, and for other Negro peculiarities', were 'savage and morally degraded'; the 'most civilized races', on the other hand, like the Mandings, 'have, as far as form is concerned, nearly European countenances and a corresponding configuration of the head'. Scientific thought accepted race superiority and inferiority until well into the twentieth century. Only in the past thirty or forty years has racism lost intellectual respectability.

Second, amid all the ramifications of contending schools of racist thought, there was total agreement on one essential point, summed up by P. D. Curtin:

> Whether the 'inferior races' were to be coddled and protected, exterminated, forced to labour for their 'betters', or made into permanent wards, they were undoubtedly outsiders – a kind of racial proletariat. They were forever barred both individually and collectively from high office in church and state, from important technical posts in law and medicine, and from any important voice in their own affairs . . . They were racially unfitted for 'advanced' British institutions such as representative democracy.

And third, there was an organic connection in nineteenth-century Britain between the attitude the ruling class took to the 'natives' in its colonies and the attitude it took to the poor at home. Though the Chartist movement evaporated after 1848, by the 1860s working people in Britain were once more challenging the political and economic power of those who ruled and employed them. Faced with this challenge, 'the proponents of social inequality slipped all the more readily into racial rhetoric.' 'Lesser breeds' and 'lower orders' had much in common, not least in the threat they presented to law and order. In Bernard Semmel's words,

> The English governing classes in the 1860s regarded the Irish and the non-European 'native' peoples just as they had, quite openly, regarded their own labouring classes for many centuries: as thoroughly undisciplined, with a tendency to revert to bestial behaviour, consequently requiring to be kept in order by force, and by occasional but severe

184 Anti-Racist Science Teaching

flashes of violence; vicious and sly, incapable of telling the truth, naturally lazy and unwilling to work unless under compulsion.

As V. G. Kiernan puts it, 'discontented native in the colonies, labour agitator in the mills, were the same serpent in alternate disguises. Much of the talk about the barbarism or darkness of the outer world, which it was Europe's mission to rout, was a transmuted fear of the masses at home'.

Bearing these three facts in mind, we can summarize nineteenth-century English racism under the broad headings of phrenology, teleology, evolutionism, anthropology, social Darwinism, Anglo-Saxonism, trusteeship, and vulgar racism.

PHRENOLOGY: BY THEIR BUMPS YE SHALL KNOW THEM

The pseudo-science of phrenology, which told people's characters from the contours of their skulls, served from the start as a prop to racism. Most of the leading phrenologists had large collections of human skulls from all over the world and firmly believed that there was a correlation between the shape of the head in different human varieties and their degree of civilization. In 1819 the distinguished surgeon Sir William Lawrence

SUGGESTED EXERCISES

1. Measure the length and width of the skulls shown in Fig. 22 and work out the cephalic index for each.
2. Make a simple, wooden measuring-gauge, as shown in Fig. 26. For the ruler,

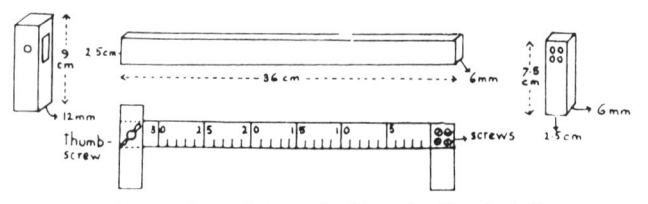

FIG. 26. Parts of a Gauge for Measuring Cephalic Indices

Anthropology.—Craniometry.

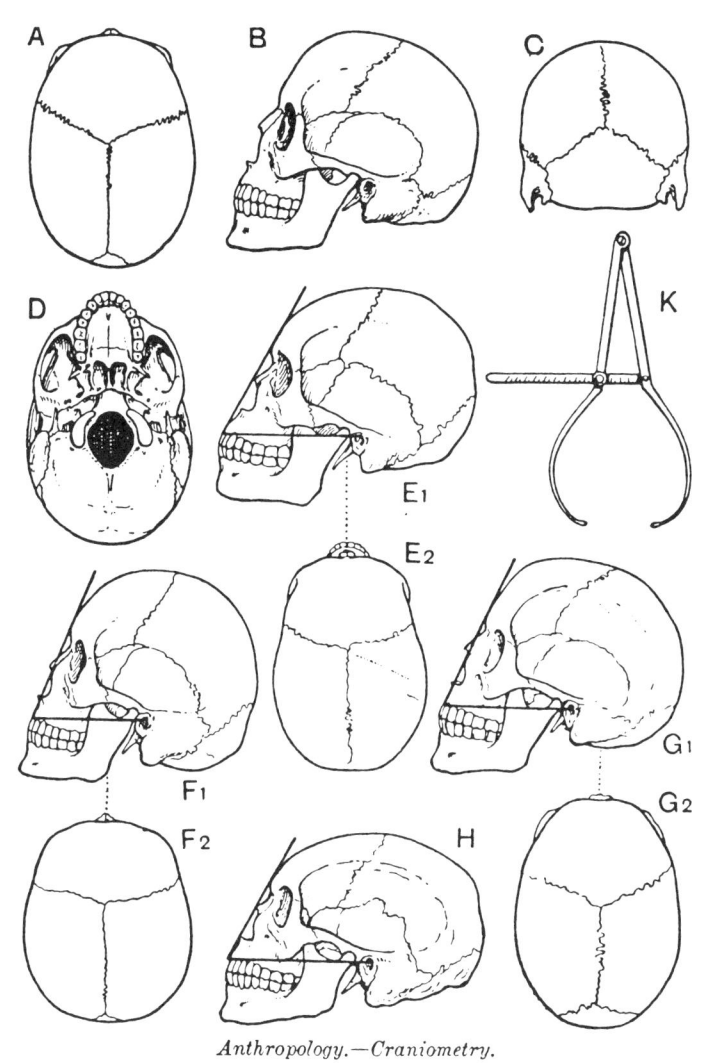

A, Norma verticalis. B, Norma lateralis. C, Norma occipitalis. D, Norma inferior. E1, E2, Dolichocephalic type (negro). F1, F2, Brachycephalic type (Chinese). G1, G2, Mesaticephalic type (European). H, Prognathous type (Australian). K, Callipers.

was using phrenology to show that race and culture were connected.

The Edinburgh lawyer George Combe, chief popularizer of phrenology in Britain, firmly believed that it could be applied to the study of race. Africans' skulls, he said, were inferior to those of Europeans. Although they showed a high development of Philoprogenitiveness, Concentrativeness, Veneration, and Hope, they lacked Conscientiousness, Cautiousness, Ideality, and Reflection. Blacks were not unfit for free labour – some as operative mechanics, others as clerks, others as 'mere labourers'.

By the 1820s phrenology was in vogue, and this gave added force to the views of W. F. Edwards, an English anthropologist who lived in France and wrote in French. For Edwards, the form and proportion of the head and face provided the crucial distinction between races. Edwards's racial interpretation of European history 'marked the beginning of a new flowering for pseudo-scientific racism', since, 'if racial interpretations of European history could be made to look "scientific", racial explanations of African culture seemed all the more plausible.'

Phrenology justified empire-building. It told the British that they were ruling over races which, unlike themselves, lacked force of character. According to Combe, before Europeans took civilization to Africa that continent exhibited 'one unbroken scene of moral and intellectual desolation'. The rich phrenologist and physician Robert Verity, an admirer of Lord Kames, predicted that 'the inferior and weaker' races would in due course become extinct and that within a hundred years Britain, its wealth, population, and intelligence, would overshadow the whole world and British civilization and language would likewise be dominant. Of all the modern nations, the English had 'a greater and more proportionate admixture of the best races': 'Eminently superior in their cerebral type, and their physical conformation, they join to these advantages the very best combination of temperament.' Here, no doubt, lay the main secret of phrenology's success. The British were already convinced of their high destiny. Phrenology told them why they were lucky and how to remain so.

TELEOLOGY: BLACKS AS BEASTS OF BURDEN

'The strongest moral force in the literature of his time.' That is how *The Cambridge History of English Literature* describes Thomas Carlyle, author of *Sartor Resartus* (1836) and *The French Revolution* (1837). He 'affirmed without fear', it adds, '... the eternal need for righteousness in the dealings of man with man'. One of Carlyle's less famous works is an *Occasional Discourse on the Nigger Question* (1853), first published under a less insulting title in *Fraser's Magazine* in 1849. From this it emerged that when white men had dealings with black men, righteousness gave place to hierarchy. As Carlyle saw it, Africans had been created inferior in order to serve their European masters. Whites were born wiser than blacks, and blacks must obey them. Carlyle trounced humanitarians 'sunk in deep froth-oceans of "Benevolence", "Fraternity", "Emancipation-principle", "Christian Philanthropy"', and thereby blinded to black people's innate stupidity and laziness. Though the radical philosopher John Stuart Mill publicly attacked Carlyle for this article, it circulated widely in pamphlet form; after all, 'it matched exactly the opinion of the Colonial Office'. In 1867, Carlyle reaffirmed his position:

> One always rather likes the Nigger; evidently a poor blockhead with good dispositions, with affections, attachments, – with a turn for Nigger Melodies, and the like: – he is the only Savage of all the coloured races that doesn't die out on sight of the White Man; but can actually live beside him, and work and increase and be merry. The Almighty Maker has appointed him to be a Servant.

The *Spectator*, quoting the essayist William Rathbone Greg in 1865, put the teleological view in a nutshell: 'The negroes are made on purpose to serve the whites, just as the black ants are made on purpose to serve the red.'

A variant of the teleological view, favoured by the medical profession, held that blacks were capable, whites incapable, of working in the tropics. Since the resources of the tropics had been put there for the whole of humanity to enjoy, they must be exploited by the labour – forced, if need be – of those capable of working there. Extreme supporters of this view went so far as to suggest that emancipation had failed and slavery should be brought back in the West Indies; Disraeli proposed this obliquely in a Commons speech in 1846.

One British colonial administrator who supported the teleological view was the explorer Sir Harry Johnston, who served as commissioner for South Central Africa from 1891 to 1896 and special commissioner for Uganda from 1899 to 1901. He wrote in 1899 that Africans, with few exceptions, were the natural servants of other races: 'The negro in general is a born slave', possessing great physical strength, docility, cheerfulness, a short memory for sorrows and cruelties, gratitude for kindness, and ability to 'toil hard under the hot sun and in the unhealthy climates of the torrid zone'; 'provided he is well fed, he is easily made happy'. The same mythology was expressed in the same period by the respected classical scholar and humanitarian Gilbert Murray, who write in 1900:

> There is in the world a hierarchy of races ... those nations which eat more, claim more, and get higher wages, will direct and rule the others, and the lower work of the world will tend in the long-run to be done by the lower breeds of men. This much we of the ruling colour will no doubt accept as obvious.

EVOLUTIONISM: THE ROAD TO EXTINCTION

Some white Americans argued in the eighteenth century that the extinction of American 'Indians' was nature's way of making room for a higher race. This evolutionary racism was imported into Britain in the 1830s and was soon dominating discussions about the proper 'native policy' for Australia, Canada, New Zealand and South Africa. In 1841 Dr Thomas Arnold, headmaster of Rugby, advanced a racial interpretation of European history. He thought the final stage of history had been reached; races unable to absorb European culture would dwindle away and, in the end, become extinct.

Pseudo-Scientific Racism 189

But suppose – just suppose – the dark-skinned races fought back? The outcome of the race struggle might not, after all, be a foregone conclusion. This chilling prospect was a central preoccupation of the Scottish anatomist Dr Robert Knox, 'one of the key figures in the general Western movement towards a dogmatic pseudo-scientific racism'. Knox's career as surgeon had been ruined by his connection with the sordid Burke and Hare body-snatching scandal. He was mobbed and burnt in effigy. He then turned to the 'science' of 'transcendental anatomy', and his popular lectures on this subject were published in 1850 as *The Races of Men*. Here is the key passage, blending racism, belligerence, and dreams of empire in equal measure:

> Look at the Negro, so well known to you, and say, need I describe him? Is he shaped like any white person? Is the anatomy of his frame, of his muscles, or organs like ours? Does he walk like us, think like us, act like us? Not in the least. What an innate hatred the Saxon has for him, and how I have laughed at the mock philanthropy of England! . . . it is a painful topic; and yet this despised race drove the warlike French from St. Domingo [i.e. Haiti], and the issue of a struggle with them in Jamaica might be doubtful. But come it will, and then the courage of the Negro will be tried against England . . . With one thousand white men all the blacks of St. Domingo could be defeated in a single action. This is my opinion of the dark races.
>
> Can the black races become civilized? I should say not . . .
>
> By ascending the Senegal cautiously and rapidly . . . a thousand brave men on horseback might seize and hold Central Africa to the north of the tropic; the Celtic race, will, no doubt, attempt this some day. On the other hand, accident has prepared the way for a speedy occupation of Africa to the south of the equator by the Saxon race, the Anglo-Saxon.

This bold armchair strategist sought to prove that 'race is everything: literature, science, art – in a word, civilization, depends on it.' A generation after Auschwitz, it is hard not to shudder when one reads his gloating vision of the outcome of racial conflict: 'What signify these dark races to us? Who cares particularly for the Negro, or the Hottentot, or the Kaffir? These latter have proved a very troublesome race, and the sooner they are put out of the way the better.' The dark races were simply animals: 'Destined by the nature of their race, to run, like all other animals, a certain limited course of existence, it matters little how their extinction is brought about.'

ANTHROPOLOGY: KILLING BLACKS NO MURDER

The phrenologists were not the only nineteenth-century 'scientists' fascinated by human skulls. But the Philadelphia physician and palaeontologist Samuel George Morton, who collected 837 of them – his collection, the world's biggest, was called the 'American Golgotha' – was interested in their capacity, not their contours. He spent a lot of time in the 1840s filling his skulls with material that would pack closely – white pepper seeds or shot pellets – and then measuring how much he had poured into each one. All this was far from easy. It was hard to devise a uniform method of closing the openings in the skulls, and hard to decide just when to stop pouring. All the same, convinced that the larger the cranium the greater the intelligence of the skull's former owner, Dr Morton persisted in his attempts to measure 'cranial capacity' and find a relationship between that and race. While confessing that there were probably errors of measurement in his tables, he found that the English skulls in his collection had an average cranial capacity of 96 cubic inches; white Americans and Germans averaged 90; Africans averaged 85, Chinese 82, and Indians 80. Since the skulls in the top three categories were nearly all from whites hanged as felons, 'it would have been just as logical to conclude that a large head indicated criminal tendencies.'

It was 'research' of this calibre on which James Hunt, an expert in the treatment of stammers, based his theory of innate differences between black people and white people. Hunt founded the Anthropological Society of London in 1863. Its vice-president was Richard Burton, who was to be found at the society's meetings 'airing his distaste for negroes, and rejoicing in the rising value of phallic specimens among European collectors'. Rajah Sir James Brooke of Sarawak and Governor Edward Eyre of Jamaica were members, and so were the poet Swinburne and Frederic William Farrar, author of the edifying school story *Eric, or Little by Little* (1858) and afterwards dean of Canterbury.

Hunt's doctrine, was set forth in his annual presidential addresses to the society, was much influenced by Dr Knox, to whose derivative teachings he himself added little that was new. Here is Hunt's own helpful summary of his 'general deductions':

> 1. That there is as good reason for classifying the Negro as a distinct species from the European, as there is for making the ass a distinct species from the zebra: and if, in classification, we take intelligence into consideration, there is a far greater difference between the Negro and European than between the gorilla and chimpanzee. 2. That the analogies are far more numerous between the Negro and the ape, than between the European and the ape. 3. That the Negro is inferior intellectually to the European. 4. That the Negro becomes more humanized when in his natural subordination to the European than under any other circumstances. 5. That the Negro race can only be humanized and civilized by Europeans. 6. That European civilization is not suited to the Negro's requirements or character.

What is really important here is the political content, which Hunt was later to amplify. As long as Britain possessed an empire, he argued, it was essential to understand the practical importance of race distinctions, because of 'the absolute impossibility of applying the civilization and laws of one race to another race of man essentially distinct'.

The political role of the Anthropological Society of London, and of its president's brand of racism, became quite clear in 1865 when one of its members, Governor Eyre of Jamaica, reacted to a rebellion of black farmers at Morant Bay with a degree of ruthlessness unusual even in the nineteenth century. He declared martial law and his troops went on a murderous thirty-day rampage, killing 439 black people, flogging at least 600 others (some were flogged before being put to death, and some were flogged with a cat among whose lashes were interwoven lengths of piano wire), dashing out children's brains, ripping open the bellies of pregnant women, and burning over 1,000 homes of suspected rebels. Public opinion in Britain was polarized, and bitter feelings were aroused on both sides. The Jamaica Committee, called by Thomas Carlyle 'a small loud group . . . of Nigger-Philanthropists, barking furiously in the gutter', was led by John Stuart Mill, Thomas Huxley, and Herbert Spencer, and its supporters included Charles Darwin and Leslie Stephen. At first it sought merely a thorough investigation and Eyre's recall, but after a whitewashing Royal Commission report it demanded the governor's prosecution for murder; at one working-class meeting he was burnt in effigy. On the other side, an Eyre Defence Committee was set up, supported by Carlyle, Ruskin, Dickens, Tennyson, Matthew Arnold, and Charles Kingsley, who thought the governor ought to be given a seat in the House of Lords. They saw him, indeed, as saviour of the West Indies and sought to raise £10,000 for his legal expenses. The Anthropological Society rallied behind its controversial member and drew some sharp political conclusions from the affair. Hunt told the 1866 annual meeting that

> we anthropologists have looked on, with intense admiration, at the conduct of Governor Eyre as that of a man of whom England ought to be (and some day will be) justly proud. The merest novice in the study of race-characteristics ought to know that we English can only successfully rule either Jamaica, New Zealand, the Cape, China, or India, by such men as Governor Eyre.

As a public service, the society invited Commander Bedford Pim, a retired naval officer, to read a paper on 'The Negro and Jamaica'. An audience of upper-class Englishmen heard Pim say that the black man in Africa and the New World was 'little better than a brute,' – in mental power a child, in ferocity a tiger, in moral degradation sunk to the lowest depths'. Slavery had rescued 'a decidedly inferior race' from a state of barbarism scarcely human. Pim supported the governor's prompt action and concluded that only through the study of 'anthropological science' could statesmen learn the true art of governing alien races. Of special interest is Pim's comparison between the discontented black people of Jamaica and the lower orders in Britain: 'We do not admit of equality even amongst our own race, . . . and to suppose that two alien races can compose a political unity is simply ridiculous. One section *must* govern the other.' Behind racism lurked the spectre of working-class rebellion in Britain.

Commander Pim's address was greeted with loud cheers and a unanimous vote of thanks, after which speaker after speaker from the floor gave advice on the technique of governing alien races. One advocated killing 'savages' as a 'philanthropic principle': when trouble broke out there might be 'mercy in a massacre'. After almost half a century, the spirit of Peterloo lived on.

Acquitted, as the *Spectator* put it, 'because his error of judgement involves only negro blood', Governor Eyre was retired on a pension.

It is hardly necessary to add that Hunt and the Anthropological Society supported the Confederacy in the American Civil War. And after that war ended they reprinted American material calling for the return of slavery in the south as the only condition under which black people would do any productive work.

SOCIAL DARWINISM: SURVIVAL OF THE FAIREST

Charles Darwin's theory of evolution proved that Europeans were related to Africans and that all human beings were related to apes. Thus it pulled the rug from under the polygenists' feet and made the whole long debate between monogenists and polygenists irrelevant. At the same time Darwinism furnished a new rationale for almost all the old beliefs about racial superiority and inferiority. Nineteenth-century sociologists assumed that when they were studying human society they were studying innate racial characteristics at the same time. White skin and 'Anglo-Saxon' civilization were seen as the culmination of the evolutionary process. So the application of Darwin's theories to human society 'had a more pervasive influence in spreading racist assumptions than the comparative anatomy of the anthropologists.'

Social Darwinism had been anticipated by Herbert Spencer, a philosopher much influenced by phrenology. It was Spencer, in fact, who coined the phrase 'survival of the fittest'. He pointed out in 1851 that the 'purifying process' by which animals killed off the sick, deformed, and old was at work in human society too, thanks to 'the decrees of a large, far-seeing benevolence'. And when Darwin published his *Origin of Species* eight years later it was Spencer who led the way in systematically applying Darwin's ideas to sociology and ethics.

Spencer had plans to study human psychological development, rather as Darwin had studied biological evolution. To understand the minds of 'primitive' races, 'civilized' races should look at the minds of their own children. The minds of 'primitive' races had similar limitations, but their childhood of intellect was permanent. Dominant races overran 'inferior' races because they had a greater 'mental mass', which showed itself in greater energy.

In many ways, however, Spencer was not a typical social Darwinian. Other believers in the natural law that the strong must devour the weak went much further in drawing racist conclusions from that law. The economist Walter Bagehot, who wrote *The English Constitution* (1867) and for whom deference to leaders was the essence of parliamentary government, argued that the strongest nations tended to conquer the weaker – 'and in certain marked peculiarities the strongest tend to be the best'. *Social Evolution* (1894) by Benjamin Kidd, a minor civil servant, made its author famous and sold 250,000 copies. It caught the popular imagination with its praise of the 'vigorous and virile' Anglo-Saxon race, in mere contrast with whom the weaker, 'inferior' races tended to die off. 'Feeble races are being wiped off the earth', wrote 'A Biologist' in the *Saturday Review* in 1896.

John Arthur Roebuck, MP for Sheffield (who made himself unpopular with working men by calling them spendthrifts and wife-beaters) told the Commons in 1862 that in New Zealand 'the Englishmen would destroy the Maori, and the sooner the Maori was destroyed the better'.

Darwin's cousin Sir Francis Galton, founder of the 'science' of eugenics, believed that 'the average intellectual standard of the negro race is some two grades below our own'. A 'very large' number of black people were 'half-witted'. Other 'inferior populations', too, were congenitally defective; Galton took a specially low view of black Australians and Spaniards. Michael Banton points out that 'the acceptance by a man of Galton's intellect and eminence of the thesis that different races could be distinguished and compared with one another, and his use of a mock statistical technique to this end, must have assisted considerably the propagation of racist theories'. Galton's pupil Karl Pearson, professor at London University and Fellow of the Royal Society, saw colonialism as a means of preparing 'a *reserve of brain and physique*' for times of national crisis: 'Such a reserve can always be formed by filling up with men of our own kith and kin the waste lands of the earth, even at the expense of an inferior race of inhabitants.' From a genetic standpoint, the black race was 'poor stock', and struggle against 'inferior races' was the way to keep a nation up to a high pitch of efficiency:

> History shows . . . one way, and one way only, in which a high state of civilization has been produced, namely, the struggle of race with race, and the survival of the physically and mentally fitter race . . .
> This dependence of progress on the survival of the fitter race . . . gives the struggle for existence its redeeming features.

Exterminated 'inferior races', Pearson added, were 'the stepping-stones on which mankind has arisen to the higher intellectual and deeper emotional life of today'.

It should be borne in mind that racial extinction was not just a matter of theory. The black people of Tasmania did not long survive the invasion of their island by the dominant race. They were hunted down without mercy. The last of them died in 1869. And racist ideology justified genocide. Social Darwinism taught white people that the Tasmanians were their brothers and sisters. It also taught them that the extermination of those brothers and sisters was an inevitable part of the struggle for existence, in which their own 'superior' race alone was destined to survive.

CONSENSUS RACISM: ENGLISHMEN, FOREIGNERS, AND NIGGERS

Hardly any British writers came out openly against racism in the last quarter of the nineteenth century. One who did was the orientalist and philologist Robert Needham Cust. 'The common form description of an African', he wrote in 1883, 'is that he is cruel, dirty, superstitious, selfish, a cannibal, and addicted to fetishism, human sacrifices, sorcery, and slave-dealing, besides being a drunkard, polygamist, and neglector of domestic ties, a liar and a cheat.' And he added: 'How different is the impression gained from an extensive consideration of the whole subject!'

Few bothered to give the subject extensive consideration. And Cust's catalogue reflects pretty faithfully the extent to which racist ideology had soaked into his unreflective countrymen's minds. Not that all educated people were out-and-out rabid racists. But 'the vast majority of the educated public appears to have accepted at least some aspect of the new racial doctrine, if only as a vague feeling that science supported the common xenophobic prejudice.'

Race prejudice was an acceptable subject for a humorous essay. The writer Charles Lamb, confessing himself to be a 'bundle of prejudices', wrote that while he had felt 'yearnings of tenderness' towards some black faces – 'or rather masks' – that had looked kindly on him in casual encounters in the streets, 'I should not like to associate with them – to share my meals and my good-nights with them – because they are black.' Lamb had similar feelings about Jews, Scotsmen, and Quakers, and his account of these feelings was published as one of his

Essays of Elia. Lamb was no racist. But pseudo-scientific racism both encouraged and fed off the kind of prejudice that afflicted him.

The flood-tide of racism never completely submerged the image of the black as 'man and brother'. Though there is more than a trace of the 'noble savage' in some of the invocations of this image, it was kept alive by three distinct traditions: humanitarian abolitionism; radicalism; and working-class solidarity. Yet the strength of these traditions should not be exaggerated. 'It would be hard to overemphasize the bias of British attitudes in the nineteenth century towards people with dark skins.' Only a minority of any social class, or at any level of education, would have raised strong objections to the 'common form description' as summarized by Cust.

The racism whose rise is here outlined has been central to the experience of black people in Britain for the past two hundred years. Racist beliefs have led to various kinds of racist behaviour on the part of many white people in Britain, including white people in authority. But black people in Britain have always asserted their humanity, dignity, and individuality in the teeth of racist beliefs and practices.

(This article is an abridged excerpt from Peter Fryer, *Staying Power: The History of Black People in Britain*, Pluto, 1984, which includes extensive references to historical documents.)

ARE THE RACES DIFFERENT?

RICHARD LEWONTIN

Racism claims that there are major inherited differences in temperament, mental abilities, energy, and so on between human groups, even though no evidence exists for such inherited differences. Racism draws credibility from what seem to be obvious differences in some physical traits like colour, hair form, or facial features. 'After all,' it is argued, 'races differ so markedly in such inherited physical traits, so isn't it reasonable that they would differ in mental ones as well?' To understand the real situation we need to look at what is really known about genetic differences between people and to examine the very concept of 'race' itself.

RACE IS ONLY SKIN-DEEP

In the nineteenth century and before, 'race' was a fuzzy concept that included many kinds of relationships. Sometimes it meant the whole species as 'the human race'; sometimes a nation or tribe as 'the race of Englishmen'; and sometimes merely a family, as 'He is the last of his race.' All that held these notions together was the idea that members of a 'race' were somehow related by ties of kinship and that their shared characteristics were somehow passed from generation to generation.

Beginning in the middle of the nineteenth century, with the popularity of Darwin's theory of evolution, biologists began to use the concept of 'race' in a different way. It simply came to mean 'kind', an identifiably different form of organism within a species. So there were light-bellied and dark-bellied 'races' of mice or banded or unbanded shell 'races' of snails. But defining 'races' simply as observable kinds produced two curious situations. First, members of different 'races' often existed side by side within a population. There might be twenty-five different 'races' of beetles, all members of the same species, living side by side in the same local population. Second, brothers and sisters might be members of two different races, since the characters that differentiated races were sometimes influenced by alternative forms of a single gene. So, a female mouse of the light-bellied 'race' could produce offspring of both light-bellied and dark-bellied races, depending on her mate. Obviously there was no limit to the number of 'races' that could be described within a species, depending on the whim of the observer.

Around 1940 biologists, under the influence of discoveries in population genetics, made a major change in their understanding of race. Experiments on the genetics of organisms taken from natural populations made it clear that there was a great deal of genetic variation between individuals even in the same family, not to speak of the same population. It was discovered that many of the 'races' of animals previously described and named were simply alternative hereditary forms that could appear within a family. Different local geographic populations did not differ from each other absolutely, but only in the relative frequency of different characters. For example, in human blood groups, some individuals were type A, some type B, some AB, and some O. No population was exclusively of one blood type. The difference between African, Asian, and European populations was only in the proportion of the four kinds.

These findings led to the concept of 'geographical race', as a population of varying individuals, freely mating among each other, but different in average proportions of various genes from other populations. Any local random-breeding population that was even slightly different in proportion of different gene forms from other populations was a geographical race. This new view of race had two powerful effects. First, no individual could be regarded as a 'typical' member of a race. Older textbooks of anthropology would often show photographs of 'typical' Australian aborigines, tropical Africans,

Ratcial differences?

and Japanese, listing as many as fifty or a hundred 'races', each with its typical example. Once it was recognized that every population was highly variable and different largely in average proportions of different forms from other populations, the concept of the type specimen became meaningless.

The second consequence of the new view of race was that since every population differs slightly from every other one on the average, all local interbreeding populations are 'races', so race really loses its significance as a concept. The Kikuyu of East Africa differ from the Japanese in gene frequencies, but they also differ from their neighbours, the Masai, and although the extent of the differences might be less in one case than in the other, it is only a matter of degree. This means that the *social* and *historical* definitions of race that put the two East African tribes in the same 'race', but put the Japanese in a different 'race', were purely arbitrary. How much difference in the frequencies of A, B, AB, and O blood groups does one require before deciding that it is large enough to declare two local populations are in separate 'races'?

ALL PEOPLE LOOK ALIKE

In ordinary parlance we still speak of Africans as one race, Europeans as another, Asians as another. And this distinction corresponds to our everyday sensory impressions. No one would mistake a Masai for a Japanese, or either for a Finn. Despite variation from individual to individual within these groups, the differences between groups in skin colour, hair form, and some facial features makes them clearly different. Racism takes these evident differences and claims that they demonstrate major genetic separation between 'races'. Is there any truth in this assertion?

We must remember that we are conditioned to observe precisely those features and that our ability to distinguish individuals as opposed to types is an artefact of our upbringing. We have no difficulty at all in telling apart individuals in our own group, but 'they' all look alike. Once, in upper Egypt, my wife was approached by an Egyptian who began a lively conversation with her under the impression that he knew her. After she repeatedly protested that he was mistaken, he apologized, saying, in effect, 'I'm sorry, but all you European women look alike.'

SUPERIORITY IS IN THE EYES OF THE BEHOLDER

If we could look at a random sample of different genes, not biased by our socialization, how much difference would there be between major geographical groups, say between Africans and Australian aborigines, as opposed to the differences between individuals within these groups? It is, in fact, possible to answer that question.

During the last forty years, using the techniques of immunology and of protein chemistry, it has been possible to identify a large number of human genes that code for specific enzymes and other proteins. Very large numbers of individuals from all over the world have been tested to determine their genetic constitution with respect to such proteins, since only a small sample of blood is needed to make these determinations. About 150 different genetically coded proteins have been examined, and the results are very illuminating for our understanding of human genetic variation.

It turns out that 75 per cent of the different kinds of proteins are identical in all individuals tested from whatever population, with the exception of an occasional rare mutation. These so-called *monomorphic* proteins are common to all human beings of all races, and the species is essentially uniform with respect to the genes that code them. The other 25 per cent are *polymorphic* proteins. That is, there exist two or more alternative forms of the protein, coded by alternative forms of a gene, that are reasonably common in our species. We can use these polymorphic genes to ask how much difference there is between populations, as compared with the difference between individuals within populations.

An example of a highly polymorphic gene is the one that determines the ABO blood type. There are three alternative forms of the gene which we will symbolize by A, B, and O, and every population in the world is characterized by some particular mixture of the three. For example, Belgians have about 26 per cent A, 6 per cent B, and the remaining 68 per cent is O. Among Pygmies of the Congo, the proportions are 23 per cent A, 22 per cent B, and 55 per cent O. The frequencies can be depicted as a triangular diagram, as shown in Figure 1. Each point represents a population, and the proportion of each gene form can be read as the perpendicular distance from the point to the appropriate side of the triangle. As the figure shows, all human populations are clustered fairly close together in one part of the frequency space. For example, there are no populations with very high A and very low B and O (lower right-hand corner). The figure also shows that populations that belong to what we call major 'races' in our everyday usage do not cluster together. The dashed lines have been put around populations that are similar in ABO frequencies, but these do not mark off racial groups. For example, the cluster made up of populations 2, 8, 10, 13, and 20 include an African, three Asian, and one European population.

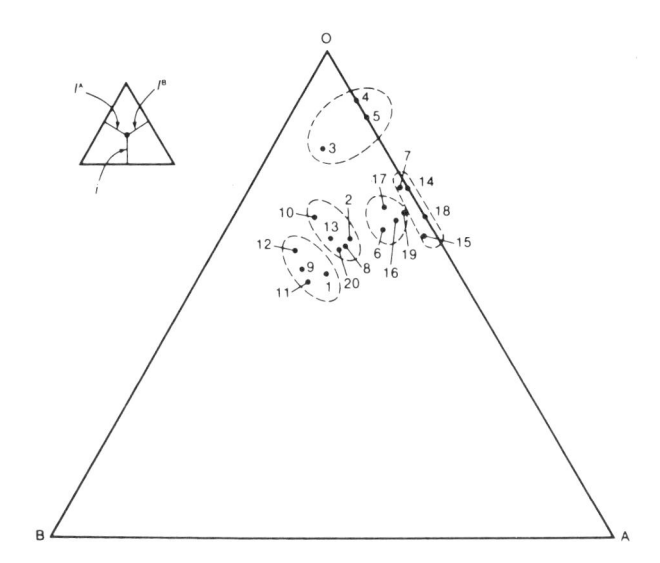

Figure 1. Triallelic diagram of the ABO blood-group allele frequencies for human populations. Each represents a population; the perpendicular distances from the point to the sides represents the allele frequencies as indicated in the small triangle. Populations 1–3 are Africans, 4–7 are American Indians, 8–13 are Asians, 14–15 are Australian aborigines, and 16–20 are Europeans. Dashed lines enclose arbitrary classes with similar gene frequencies, which do not correspond to the 'racial' classes.

A major finding from the study of such polymorphic genes is that none of these genes perfectly discriminates one 'racial' group from another. That is, there is no gene known that is 100 per cent of one form in one race and 100 per cent of a different form in some other race. Reciprocally, some genes that are very variable from individual to individual show no average difference at all between major races. (See Table 1.)

Rather than picking out the genes that are the most different or the most similar between groups, what do we see if we pick genes at random? Table 2 shows the outcome of such a random sample. Seven enzymes known to be polymorphic were tested

Table 1: Examples of extreme differentiation and close similarity in blood-group allele frequencies in three racial groups

Gene	Allele	Caucasoid	Population Negroid	Mongoloid
Duffy	Fy	0.0300	0.9393	0.0985
	Fy^a	0.4208	0.0607	0.9015
	Fy^b	0.5492	0.0000	0.0000
Rhesus	R_0	0.0186	0.7395	0.0409
	R_1	0.4036	0.0256	0.7591
	R_2	0.1670	0.0427	0.1951
	r	0.3820	0.1184	0.0049
	r'	0.0049	0.0707	0.0000
	Others	0.0239	0.0021	0.0000
P	P_1	0.5161	0.8911	0.1677
	P_2	0.4839	0.1089	0.8323
Auberger	Au^a	0.6213	0.6419	—
	Au	0.3787	0.3581	—
Xg	Xg^a	0.67	0.55	0.54
	Xg	0.33	0.45	0.46
Secretor	Se	0.5233	0.5727	—
	Se	0.4767	0.4273	—

SOURCE: From a summary provided in L. L. Cavalli-Storza and W. F. Bodmer, *The Genetics of Human Populations*, W. H. Freeman and Company, 1971, pp. 724–31. See this source for information on other loci and for data sources.

Table 1 shows the three polymorphic genes that are most different between 'races' and the three that are most similar among the 'races'. The first column gives the name of the protein or blood group and the second column gives the symbols of the alternative forms (alleles) of the gene that is varying. As the table shows, there are big differences in relative frequencies of the allele of the Duffy, Rhesus, and P blood groups from 'race' to 'race', and there may be an allele like FY^b that is found only in one group, but no group is 'pure' for any genes. In contrast, the Auberger, Xg, and Secretor proteins are very polymorphic within each 'race', but the difference between groups is very small. It must be remembered that 75% of known genes in humans do not vary at all, but are totally monomorphic throughout the species.

Table 2: Allelic frequencies at seven polymorphic loci in Europeans and black Africans

Locus	Europeans Allele 1	Allele 2	Allele 3	Africans Allele 1	Allele 2	Allele 3
Red-cell acid phosphatase	0.36	0.60	0.04	0.17	0.83	0.00
Phosphoglucomutase-1	0.77	0.23	0.00	0.79	0.21	0.00
Phosphoglucomutase-3	0.74	0.26	0.00	0.37	0.63	0.00
Adenylate kinase	0.95	0.05	0.00	1.00	0.00	0.00
Peptidase A	0.76	0.00	0.24	0.90	0.10	0.00
Peptidase D	0.99	0.01	0.00	0.95	0.03	0.02
Adenosine deaminase	0.94	0.06	0.00	0.97	0.03	0.00

SOURCE: R. C. Lewontin, *The Genetic Basis of Evolutionary Change*, Columbia University Press, 1974. Adapted from H. Harris, *The Principles of Human Biochemical Genetics*, North Holland, Amsterdam and London, 1970.

in a group of Europeans and Africans (actually black Londoners who had come from West Africa, and white Londoners). In this random sample of genes there is a remarkable similarity between groups. With the exception of phosphoglucomutase-3, for which there is a reversal between groups, the most common form of each gene in Africans is the same form as for the Europeans, and the proportions themselves are very close. Such a result would lead us to conclude that the genetic difference between blacks and whites is negligible as compared with the polymorphism within each group.

The kind of question asked in Table 2 can in fact be asked in a very general way for large numbers of populations for about twenty genes that have been widely studied all over the world. Suppose we measure the variation among humans for some particular gene by the probability that a gene taken from one individual is a different alternative form (allele) from that taken from another individual at random from the human species as a whole. We can then ask how much less variation there would be if we chose the two individuals from the same 'race'. The difference between the variation over the whole species and the variation within a 'race' would measure the proportion of all human variation that is accounted for by racial differences. In like manner, we could ask how much of the variation within a 'race' is accounted for by differences between tribes or nations that belong to the same 'race', as opposed to the variation between individuals within the same tribe or nation. In this way we can divide the totality of human genetic variation in a portion between individuals within populations, between local populations within major 'races', and between major 'races'. That calculation has been carried out independently by three different groups of geneticists using slightly different data and somewhat different statistical methods, but with the identical result. Of all human genetic variation known for enzymes and other proteins, where it has been possible to actually count up the frequencies of different forms of the genes and so get an objective estimate of genetic variation, 85 per cent turns out to be between individuals within the same local population, tribe, or nation. A further 8 per cent is between tribes or nations within a major 'race', and the remaining 7 per cent is between major 'races'. That means that the genetic variation between one Spaniard and another, or between one Masai and another, is 85 per cent of all human genetic variation, while only 15 per cent is accounted for by breaking people up into groups. If everyone on earth became extinct except for the Kikuyu of East Africa, about 85 per cent of all human variability would still be present in the reconstituted species. A few gene forms would be lost, like the FY^b allele of the Duffy blood group that is known only in American Indians, but little else would be changed.

WHO'S WHO?

The reader will have noticed that to carry out the calculation of partitioning variation between 'races', some method must have been used for assigning each nation or tribe to a 'race'. The problem of what one means by a 'race' comes out forcibly when making such assignments. Are the Hungarians Europeans? They certainly *look* like Europeans, yet they (like the Finns) speak a language that is totally unrelated to European languages and belongs to the Turkic family of languages from Central Asia. And what about the modern-day Turks? Are they Europeans, or should they be lumped with the Mongoloids? And then there are the Urdu- and Hindi-speaking people of India. They are the descendants of a mixture of Aryan invaders from the north, the Persians from the west, and Vedic tribes of the Indian subcontinent. One solution is to make them a separate race. Even the Australian aborigines, who have often been put to one side as a separate race, mixed with Papuans and with Polynesian immigrants from the Pacific well before Europeans arrived. No group is more hybrid in its origin than the present-day Europeans, who are a mixture of Huns, Ostrogoths, and Vandals from the east, Arabs from the south, and Indo-Europeans from the Caucasus. In practice, 'racial' categories are established that correspond to major skin colour groups, and all the borderline cases are distributed among these or made into new races according to the whim of the scientist. But it turns out not to matter much how the groups are assigned because the differences between major 'racial' categories, no matter how defined, turn out to be small.

The result of the study of genetic variation is in sharp contrast with the everyday impression that major 'races' are well differentiated. Clearly, those superficial differences in hair form, skin colour, and facial features that are used to distinguish 'races' from each other are not typical of human genes in general. Human 'racial' differentiation is, indeed, only skin-deep. Any use of racial categories must take its justification from some source other than biology. The remarkable feature of human evolution and history has been the very small degree of divergence between geographical populations as compared with the genetic variation among individuals.

Assessment

INTRODUCTION

Most of this book emphasizes the need to change and expose the racist ideology in curriculum content, and provide anti-racist education. However, it is not enough to insert such changes into the predominant methods of assessing students and exercising teacher authority – such as using the prospect of exams to motivate and control students. That would only reduce 'anti-racism' to yet another imposed authority and perpetuate the 'low achiever' labelling of many students.

We would argue that the notions of underachievement and educational disadvantage need to be examined afresh. Their current use defines pupils and their families as the problem. We need to appropriate these terms as conceptual tools which focus attention not on pupils and their families but on the education system and the mechanisms by which it works to 'disadvantage' students. 'Disadvantage' needs to be made into an active verb. The ILEA goes some way towards reconceptualizing the notion of disadvantage in criticizing the government's approach to underachievement, which defines the problem simply as 'the bottom 40 per cent outside the target area of the examination system', for whom separate provision is supposedly needed. On the contrary, ILEA's leader argues:

> The current examination system does not simply disadvantage the 'bottom 40 per cent'; it is an ineffective and inadequate method of recording the achievements of the 60 per cent who in some sense survive within it. It is not simply

educationally undesirable that attitudes and values of educational institutions should be based on assumptions of differential provision in relation to gender, class or ethnic origin; the quality of the education itself suffers too. (ILEA No. 1, 1983, p. 5.)

Unfortunately the assumption is widespread that 'ethnic minority' pupils, especially those of West Indian origin, suffer from a 'racial disadvantage' on account of special 'ethnic' needs which go unmet because of teacher racism, covert and overt (Sivanandan, 1984, p. 23). Often equating special needs with disabilities, such approaches reduce underachievement to a dual problem of 'ethnicity' on the one hand and teacher attitudes on the other – both separated from the class role of the school system. This approach – which underpins the recommendations of the 'Swann Report' – leads to demands for a 'multicultural curriculum', sometimes even for differential provision for black and white pupils, and to the furores that these demands bring in their wake. Multiculturalism fails to address the structural nature of racism. By locating the problem in black pupils and their families, it serves further to divide pupils and to divert attention away from the common oppression of black and white working-class pupils.

We argue in this section that the assessment procedures adopted in schools must become the focus of critical attention if the notion of anti-racist education is to mean anything at all. Assessment procedures discriminate not simply against black pupils, but against white working-class students and girls. Through assessment, schooling serves to disadvantage most students, while labelling many as failures.

Failure is individualized through any assessment process currently used in schools. Through it pupils learn not only to see themselves as successes or failures, but to accept personal responsibility for their own success or failure. The process of assessment does a double job. It selects pupils for certain roles in the social and economic order, while at the same time socializing them into accepting those roles as a just reward for their 'ability'.

Thus we need to change more than teacher attitudes and curriculum content if we want to challenge racism in schooling. The system of assessment, with the pedagogy, also needs to be changed. In adopting a critical perspective on methods of ability labelling and assessment, we challenge not only racism but also sexism and class divisions, hierarchies based on class.

This collection of articles includes a collectively written critique of 'graded tests' in science, known in ILEA as GASP. The authors argue that – far from alleviating the problem – this innovation would worsen it through yet more elaborate labelling. One of the authors, Birendra Singh, extends the critique by suggesting that such innovation hijacks 'process skills' in ways that serve the wider class role of assessment. Les Levidow offers a critique of 'ability' labelling more generally, which he attributes not just to testing but also to teaching methods which intensify certain differences that students bring to school. (His article contains an extensive Bibliography for this entire section.)

GASP! A CRITIQUE

GRAZYNA BARAN, LES LEVIDOW and BIRENDRA SINGH

By now many science teachers will have heard of proposals from the Graded Assessments in Science Project. Its acronym, GASP, unintentionally expresses the unease that some of us have felt about the Project. Our criticism has been all the more difficult to formulate because GASP presents itself as a progressive reform designed to benefit the least successful pupils. And, after all, it comes from within a local authority noted for its strong statement of anti-racist policies.

But how do we judge whether GASP would really combat or perpetuate racism? What does it propose to give to pupils? What does it tell pupils that they need or should want? Whose problem does it attempt to solve? These are the questions that we would like to take up here, in the hope of generating wider debate among teachers, parents and pupils.

WHAT ACHIEVEMENT?

First of all, GASP's concern with pupil 'underachievement' in science must be seen in the wider context of proposed reforms of testing and schooling as a whole. From the state's standpoint, underachievement poses several problems – e.g. pupil indiscipline resulting from their feeling excluded, and lowered 'value for money' (teacher productivity), among others. The Hargreaves Report, *Improving Secondary Schools*, hopes to find a solution in enhanced pupil motivation, which it defines as itself a type of achievement ('aspect IV'), not directly measured by existing exams. Using progressive-sounding language, the report expresses concern that 'working-class pupils are particularly vulnerable here, since some of them, because of disadvantaged environments, come to school with already-low levels of aspect IV achievement . . .'

Therefore, the report argues, such pupils rely more on teachers and exam results for their sense of self-esteem. In order to enhance pupil motivation, the report proposes new types of 'graded tests' measuring achievement at levels lower than that of the existing exams. In other words, the proposed solution is a generalized extension of quantitative assessment – in the name of making pupils feel that they are achieving something.

As we have just seen, the Hargreaves Report in effect adopts the notorious 'deficit' model of working-class pupils, who are seen as disadvantaged by their social background rather than by the type of schooling on offer. Never questioned are the dominant criteria of 'achievement'. In our view, schools actively disadvantage working-class (and black) pupils by defining achievement in an individualistic, competitive way.

This basis of individual assessment guarantees 'failures' as well as 'successes'; indeed, only on such a basis could assessment provide a 'motivation'.

Science teaching, in particular, emphasizes the written use of a technical language that makes most students feel not only excluded but also isolated from each other; another structural aspect is the classroom 'practical' – coming as the culmination of a series of abstract concepts, yet unconnected to intuitively known problems or gadgets. This process discourages pupils from formulating proposals or questions in their own language and thus discourages them from appropriating scientific knowledge as their own, through joint practical experience. Result: underachievement, taken as evidence of low ability. This labelling process occurs every day, not just through occasional exams, for the curriculum gets structured according to criteria of 'achievement' defined by exams.

GASP AGAIN

What, then, about GASP? Its first newsletter (April 1984) begins: 'All school subjects contain their own intrinsic motivation, but the demands of examinations tend to destroy this.' Does it therefore propose abolishing exams? No chance. Presuming, like the Hargreaves Report, that pupil motivation depends upon doing well in exams, GASP likewise proposes a complex proliferation of quantitative assessment supposedly measuring hitherto-unmeasured skills. For example, teachers' concerns over learning as process become taken over with new tests designed to measure 'process' skills, as distinct from 'content' skills! (Is this kind of distinction we have had in mind when promoting 'process'?)

GASP documents themselves acknowledge that 'pupil motivation' is not their only rationale. Also essential is some new universal equivalent of achievement for those pupils who perform badly on present exams. GASP aims both 'to give *all* pupils a sense of achievement in science and so help motivation throughout their secondary schooling'; and 'to give outcomes fully acceptable to and understandable by employers and others.'

Another GASP document (Paper CC1), accepting notions of 'low ability', argues that therefore

> Pupils should be able to achieve some credits of evident value, however limited their ability; the most able should eventually receive credits which have currency equivalent to that of conventional assessments . . .
>
> There must be external validation so that the pupils' achievements have attested value to the world outside. It follows that the scheme will include external moderation of internal assessment, some internal assessments based on selection from an externally provided and standardized bank . . .

Another document goes further in portraying intensified 'skill'-testing as a progressive advance: 'The term "ability" often implies an attribute which is innate and resistant to change as in "low-ability pupils", so it might be better not to use it in connection with graded assessment.' What does all this mean?

All these features (banks, credits, currency) show that GASP is extending the model of 'ability' already imposed by the present examination system; GASP differs only in offering yet more assessment. In the name of ceasing to label pupils as 'less able', GASP proposes to replace such labels with 'skill' assessments as quantified by yet more exams. Of course this supposed solution aggravates the pupils' original problem, since it means finding new ways of motivating their active co-operation in what is effectively an oppressive, class-based, racist labelling system.

GRADING FOR WHOM?

The problem being solved here, it would seem, is social instability. GASP is likely to be used to con and incorporate the most rebellious pupils, especially Blacks, who are not merely victimized by discrimination but who actively reject the dominant criteria of 'achievement' and individual assessment. Indeed, GASP can be seen as a potential instrument for breaking down the camaraderie by which working-class pupils defend themselves from being divided up for individual assessment. As for those pupils who persist in refusing to jump through the inane hoops of GASP-inspired testing, the school authorities could more easily isolate and penalize them for blatant non-cooperation. Would not many of these be black pupils?

In particular, GASP would further subdivide the approximately 40 per cent of school leavers (and well over half of Blacks) who do not obtain GCE or CSE qualifications, yet who have often been able to hustle their way into various jobs by getting the employer to give them 'the benefit of the doubt' about their skills. While examination-certification systems have always tended to grade pupils in a manner generally suitable for employers, the present system is considered too vague, especially for the bottom 40 per cent. So GASP proposes to give more precise assessments – whose only consequence could be a more effective pigeonholing of prospective employees.

In particular, GASP proposes to replace present 'ability' labels with fragmented 'skill' labels – largely 'skills' that have been long taken for granted as universal; this would reinforce the existing division of labour along the lines of mental/manual, conceptual/practical divisions. In other words, the proposed GASP certificates would give more precise confirmation of a pupil's failure (to get even a GCE qualification). So its only effect would be to get the school leaver placed in the worst part of the capitalist division of labour, and in a way more effective for employers to exploit them.

WHOSE ASSESSMENT?

From an anti-racist standpoint, it is necessary to oppose GASP for structural reasons, irrespective of the GASP team's intended aims. However, we must also challenge the present social organization of science education. Whatever alternatives are devised to challenge it, they are bound to lend themselves less well to quantified assessment of individual pupils; indeed, such alternatives would mean not only transforming the criteria of 'achievement' but having such criteria defined collectively by the pupils themselves, so that the group is assessing both itself and the teacher. (Whether as a result the pupils do better or worse on the official exams is a separate but perhaps related question.)

As opposed to the individualistic, exam-based definition of achievement, we propose fundamentally different criteria: a continuous assessment of a co-operative group process that encourages the pupils' self-confidence in posing questions in their own intuitive language and connects that language to technical terminology through examples from their own experience, in or outside the school. Of course, some parents, especially those of 'underachieving' pupils, may see such a process as depriving their children of the certification or 'currency' needed to compete in the job market. But our proposal counterposes authentic self-confidence to a false motivation based on GASP's demeaning 'currency' that would relegate school leavers to the worst jobs or none at all.

Naturally, this short article cannot be expected to resolve the present uncertainties about the best method of teaching and assessment. However, it has shown how GASP takes those uncertainties in a regressive direction, a step towards institutionalizing the practices (especially testing, grading and streaming) that contribute to the failure of working-class and especially black pupils. How can co-operation with GASP be seen as consistent with ILEA's stated anti-racist policies?

GRADED ASSESSMENTS: HIJACKING 'PROCESS'

BIRENDRA SINGH

The present debate on assessment in science seems to arise from a concern about 'underachievement', especially a lack of motivation among many thirteen- to sixteen-year-olds. The debate is also partly a response to the complaints of some employers that the schools are failing to teach the 'right' skills to children; in the government's view (and some employers'), this failure is at least partly to blame for the high unemployment among school leavers. There is a demand for more flexible workers who will be prepared to transfer their skills, without question, to new situations.

As regards science in particular, there is a feeling that assessment methods should recognize the 'process skills' acquired in learning. There is unease that the traditional examination structures (O level, CSE and 16+) do not measure these skills. There is a related complaint that most of science teaching is 'content'-orientated and is geared to the requirements of an inadequate examination system.

In response to such concerns, educationalists have devised systems of 'graded assessments'. These are intended to recognize each student's accomplishments, however modest, and to emphasize process over content, thus enhancing students' motivation. In particular, a Graded Assessment in Science Project (GASP) has been initiated by the Inner London Education Authority (ILEA) Science Inspectorate in association with Chelsea College's Centre for Science and Mathematics Education and the University of London Examining Board.

Also, aspects of graded assessments have been incorporated into the new national exam system, the General Certificate of Secondary Examination (GCSE).

Are graded assessments in the interests of students? Will they help to improve science teaching? This article will evaluate both examples – GASP and the GCSE – in the light of the political role of educational assessment in this society.

GASP

The GASP team propose a series of graded assessments which include three dimensions: content, process skills and explorations. They hope that this system will help to improve science teaching itself as well as assessment. Ideally, teachers will begin to move away from a 'content' approach to a 'process'-orientated approach, with a strong component of investigations and problem-solving under the heading of 'explorations'. A recognition of achievement in process skills will spur motivation, especially because pupils will take the achievement tests only when 'ready' to do so. Thus, they believe, the system will be fairer, better indicating pupils' achievement.

For the GASP team, the goal is to enable every student in the 16+ group to receive a certificate of assessment issued on the basis described above. Ideally, the certificate will enhance their chances of obtaining appropriate jobs and will help employers to fit 'round pegs' into round holes. 'Surely matching pupils' abilities to the needs of the job cannot be bad . . .', the GASP team insist. In their view, the school system rightly ascertains differences in ability and labels students for the job market. These are 'desirable ends for any assessment system', as they 'consider that diversity is a central and delightful characteristic and would resist attempts to press everyone into the same mould' (GASP Team, 1984).

Their optimism notwithstanding, schools 'reveal' individual differences by setting, streaming and making most pupils feel that they are too 'thick' to pass exams – hardly a 'delightful' practice. Although the GASP team want to improve that situation, their model of achievement still reduces ability to an intrinsic characteristic and reduces diversity to an ordinal ranking of such 'ability'. Thus they limit themselves to seeking a supposedly fairer way to help allocate students to appropriate jobs, as if each student were naturally suited by ability for a particular job.

The GASP team also limit themselves by accepting that the graded tests must be consistent with the national examination system, which they do not challenge. The graded tests are meant for all pupils, but will they have the same meaning for all pupils? In practice, employers and other users of examination results will certainly compare the GASP grades with those of the new GCSE. Specifically, the GCSE Grades A, B and C will correspond with the O level Grades A to C and with CSE Grade 1. The GCSE Grades D to G will correspond with the CSE Grades 2 to 5. There has been no correspondence yet agreed between the GASP grades and the GCSE grades, though it is thought that each GASP certificate will specify one of about fifteen possible grades.

One can see that a pupil with an O level pass (GCSE Grades A to C), or even with a CSE pass (GCSE Grades D to G), will be able to cite his/her assessment grades as evidence of success. But what about those below GCSE Grade G? Will they confidently claim their success, or will their GASP certificate simply be a finer grading of their failure? In this context it is worth noting that in the 1960s the Northern Universities Joint Matriculation Board used to grade their O level candidates on a scale 1 to 10. Grades below 7 were perceived, both by pupils and others, as a confirmation of failure rather than of success.

Thus it is not sufficient to say that the present examination structures are inadequate simply because they are content-based and exclude process skills. The O levels are a filter for A levels, which in turn are filters for university entrance. Their design favours a middle-class view of achievement, based on an individualist notion of competition. The O levels contain within them the notion of intrinsic 'intelligence' or 'ability', while concealing the fact that these O levels have less to do with the intelligence or ability of a person than with the way the present society selects and defines its potential scientists. It is therefore essential to tell students that failure to pass O level is not a reflection of a pupil's intelligence/ability; rather, it shows an inability to accept the nature and scope of knowledge as 'packaged' by examination boards and presented in science classrooms.

Now let us look at the likely effects of graded tests, in certifying skills which have hitherto remained unmeasured. In the past, when there was a relative shortage of labour in the economy, firms employed youngsters even without school qualifications and provided facilities for training on the job; motivation for learning there was linked to wages and a desire to gain security through a good job. The recent effort to grade and subdivide those who remain ungraded by the CSE and O levels – and who may remain so by the new GCSE examinations – appears to be a response to the demands of a contracting economy. Further subdivisions of grading will have the effect of reducing pupils' confidence in their own ability by leading them to accept the notions of intelligence, ability and worth as defined by the assessment system, 'which remains consistent' with O level and CSE examination structures; it will facilitate a finer pigeonholing of pupils by schools and employers. Unemployment will then be explained by low GASP grades.

PROCESS SKILLS

GASP's move to de-emphasize content in favour of process and to encourage explorations is a step forward. It has been a major concern of teachers who have, over a period of time, fought to liberate science teaching from examination constraints. Indeed, to have one-third content, one-third process skills and one-third explorations based on problem-solving, with pupil control over this, would be a major advance, particularly if this were done in a way that would integrate all three components.

Yet this is not what GASP proposes. Its continuous assessment system, which is supposed to free pupils from the examination system, appears to substitute another way of holding them just as firmly. It does not have to do so, but from what is known of its progress so far, it appears to do so.

In particular, it hijacks process by:
☐ breaking up process skills and pretending that they can be isolated as measurable quantities separate from content and explorations;
☐ paying only lip service to the idea of explorations and allowing very little pupil control over them;
☐ separating out the pupils in an individualistic way; and
☐ requiring an unrealistically high amount of assessments for teachers to carry out.

It is therefore difficult to see how this version of processes can be in the interests of the pupils. The proponents have ignored one of the main reasons why science teaching is content-orientated: that is, the image of science that so many teachers have. To them science is a set of truths to be passed on. This view – coupled with their acceptance of intelligence and/or ability as an innate attribute – prevents them from exploring ways in which children can be encouraged to take an active part in the learning process. Under the pressures of the examination syllabus and the school administration, the class-

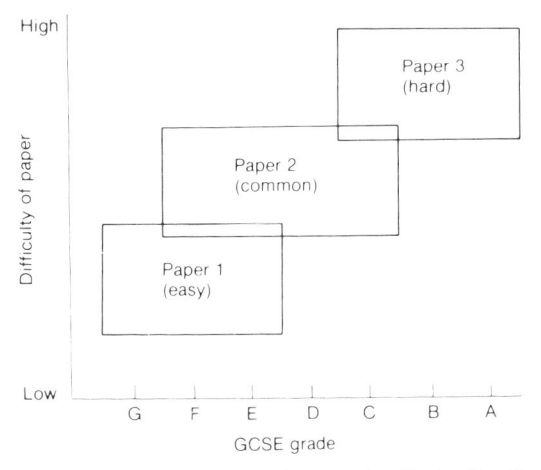

Candidates taking Papers 1 and 2 can achieve Grades G to C
Candidates taking Papers 2 and 3 can achieve Grades F to A

room teacher – who is most likely to be under-resourced, overworked and underpaid – accepts the easier option of sticking to the content of the syllabus. The examination boards do not, the school as an institution does not, and teaching as a profession has not considered seriously the social nature of ability, how it in fact depends upon the social situation (e.g. peer group), the power structure of society, and the political base of knowledge.

A science teacher, therefore, spends a considerable amount of her/his time testing, grading and sometimes 'streaming', no doubt believing that this process is somehow educationally desirable. Yet racism, sexism and classism are all supported by – and to some extent rely on – the individualistic notion of ability described earlier. Unless the notion is rejected and the damage caused by it is recognized, it will continue to dictate the way in which process skills are assessed; even the 'process' will become a testable commodity. Hence there will be no escape for pupils from the repression of the examination system. It would seem that teachers' wishes to move towards process-orientated courses are being hijacked by GASP, which presses 'process' into the mould of the present 'ability' labelling.

Like the process skills, the 'explorations' component of GASP holds out the possibility of progressive change in principle. In reality, that possibility could be realized only if pupils and teachers had control over the curriculum, if explorations and problem-solving were a major part of the curriculum, and if content, process skills and problem-solving were all integrated. But GASP minimizes such a possibility by reducing them to supposedly separable bits.

Graded tests cannot overcome classroom disaffection as long as they embody an individualist notion of innate ability and a model of fragmented knowledge, especially in a school system which makes attendance compulsory. Unfortunately, the entire assessment debate seems focused less on making science accessible than on gearing the school system still further to the present division of labour. This means pigeon-holing school leavers for underemployment and increasing unemployment, while providing the appropriate labels for lowering expectations and justifying such outcomes.

At this point it is worth quoting some relevant passages from ILEA reports:

> Whilst our education system remains unchanged it is hard to see how a major improvement in the achievement of children from working-class families could be brought about. Because our secondary curriculum is dominated by public examinations designed to mark pupils into fine grades on the basis of performance in timed tests, the gap between the social classes is likely to remain. (ILEA, *Race, Sex and Class* Booklet No. 1, p. 10)

> One reason why the relationship between social disadvantage and educational attainment is so strong within Britain is the essentially competitive system of public examinations. (ILEA *Race, Sex and Class* Booklet No. 2, p. 18)

> Pupils should be equally esteemed whatever their academic ability . . . Public examinations are an inadequate measure of educational achievement. (Hargreaves Report, 1984)

In the light of those observations, what can a science teacher do to overcome the negative aspects of the existing exams? One step would be to include in science teaching a critique of the exam system itself, so that pupils can understand why examination boards package a certain form of knowledge as the official valid knowledge. When pupils understand the true nature of the exams, they will be better equipped to define their own positions in relation to the system. Thus those who reject it, or those who fail to pass exams, have a better chance of avoiding a sense of personal failure and inadequacy.

Beyond offering students a critique of the present system, teachers need to develop alternative forms of teaching and assessment which provide all students with a sense of positive achievement. This means challenging in practice the notion of innate ability, the associated competitive mode of learning and the fragmented model of knowledge. Only in this way can teachers make science accessible by fully involving pupils in a co-operative process of learning.

GCSE

The new national examination system, the GCSE, claims to offer an improved method for 'positive assessment in science'. Criticizing former assessment methods, the GCSE Science *Guide for Teachers* states: 'The principal goal has been to produce assessment schemes through which candidates can be sorted according to their positions on a spectrum of performance. That is, examinations have been designed to "discriminate" or place candidates in rank order' (SEC, 1986).

The GCSE general criteria, however, 'require us to do more than achieve discrimination'. That is, 'all examinations must be designed in such a way as to ensure proper discrimination so that candidates across the ability range are given opportunities to demonstrate their knowledge, abilities and achievement: that is, to show what they *know*, understand and *can do*' (paragraph 16).

The GCSE Science Guide quotes the following passage, from the Cockcroft Report on the teaching of Mathematics, to show a concurrence of thinking between the GCSE and that much-acclaimed report:

> We cannot believe that it can in any way be educationally desirable that a pupil of average ability should, for the purposes of obtaining a school-leaving certificate, be required to attempt an examination paper on which he is able to obtain only one-third of the possible marks. Such a requirement, far from developing confidence, can only lead to feelings of inadequacy and failure.

The GCSE Science Guide rightly indicates that the point made in the Cockcroft Report 'seems, from the evidence, to apply with considerable force in the case of science examinations.'

Referring to the public perceptions of the grades issued by the present 16+ examinations, the above document quotes a recent survey (Goacher, 1984) in which various users of examination results were asked their views on the meanings to be attached to certain grades. Two comprehensive school teachers offered the following opinions:

Grades D and E are fails and do not deserve certification.
CSE Grade 5 is really a condemnation and not a certification.

The document asks: 'Do the ways in which we assess contribute to perceptions of this kind?'

The way we assess is linked with the way we teach. The way we perceive learning to take place, the notions of ability we have and what we consider worth knowing, that is, valid knowledge, contribute to our concept of assessment. Therefore, our assessment is not an independent activity; it is linked with the pedagogy of teaching. It is not sufficient simply to look at assessment in isolation. For the GCSE,

> the key message is that all candidates can solve problems using scientific methods. These skills are not the sole preserve of those destined to be research scientists and technologists. Candidates who achieve Grade F (equivalent to CSE 4) and those who achieve Grade A may well perceive and find manageable rather different sorts of problems. These can still be valid problems in science. Differentiated schemes of assessment are intended to present students with tasks and problems to which they can respond effectively.

So, the intention is to use 'differentiated schemes of assessment' in order 'to present candidates with differentiated tasks which build in the likelihood of success'. By so doing, the GCSE would hope to cater for many of the bottom 40 per cent pupils who have remained ungraded in the O level and CSE grading schemes.

The GCSE Science *Guide for Teachers* gives examples of written examinations which can be used to 'ensure that all candidates are presented with tasks at which they can succeed.' Secondly, it is intended that science courses leading to the GCSE will 'bring together the process (methods, procedures and skills) and the contexts and concepts of science.' To fulfil this intention, the national criteria for each of the science subjects 'provide for significant proportions of the available marks to be allocated' to the assessment of students' competence in scientific processes.

EXAM TASKS FOR SUCCESS

The proponents of GCSE hope to achieve all the aforesaid objectives by asking teachers and examiners to accept a challenge: 'to find ways of helping students to meet each objective' by ensuring that 'all candidates are presented with tasks at which they can succeed.' How well will teachers rise to this challenge? Consider the outcome of my school Phase 3 GCSE training, the final stage of training before we tackled the 'challenge' in September 1986.

I and three colleagues who teach physics at my school took up the GCSE advice and set upon the task of learning to assess the GCSE way. I must mention, first, that we are lucky to have four physics teachers in one school; secondly, we all team-teach in mixed-ability groups and get on well with each other. In addition, in our team there is one teacher who has been active in the Nuffield Physics Project since its inception and is a member of the Nuffield Consultative Committee. Most of us have both present and past examining experience; together we are highly qualified. All this – and the fact that we work as a team – should make our task at least as easy as that of any other physics teacher engaged in the same exercise.

Among many activities, we undertook two specific tasks:
1. Devise a practical test that an average pupil (GCSE Grade F = CSE Grade 4) can do, i.e. obtain at least 50 per cent of the allocated marks.
2. Look at some of the published physics syllabuses in detail.

In the first task, we *all* failed to write a test in which a Grade F pupil could obtain at least 50 per cent. Despite our sympathy with the objective of this task, we found that all our training and practice as science teachers militated against writing such a test. In other words, we are so used to writing 'tests' to make pupils fail that we found ourselves unable to write a test where most of them would pass.

In the second task we became painfully aware of the gulf

which exists between the GCSE's assessment intentions and the examining boards' interpretation of those intentions. For example, the National Criteria for Physics recommend that process skills be allocated 'not less than 40 per cent (at least 20 per cent to experimental skills)' for assessment purposes. Yet none of the examining boards' syllabuses had allocated 40 per cent for process skills; most had opted for the minimum recommended 20 per cent for experimental skills, with no allocation to any other process skills.

The marking scheme for the experimental skills, apart from being cumbersome, appeared to overemphasize testing and grading. One examining board (LEAG, 1986) identifies six assessable practical skills: observation; measurement; manipulation; design and plan; use of data; record and report. Each of those in turn is subdivided into five skill levels. This assessment procedure gives a total of $6 \times 5 = 30$ levels of grades, each worth one mark.

Who on earth would want to devise thirty levels of grading purely for the purposes of assessment? Surely the only justification must be to be able to teach in such a way that pupils become thoroughly competent in the scientific processes encapsulated in the aforementioned skills. If this pedagogical intention is to be fulfilled, then it has profound implications for time available to teachers for preparing, marking and teaching a successful GCSE course. Yet the Science Guide offers no clear advice on how the positive assessment method should be linked to positive teaching strategies.

What can we as teachers learn from our two GCSE training exercises? The lesson of our first task is that – even for an experienced, qualified team – considerable resources will be needed for training teachers in the task of positive assessment. Simple exhortations (however well meant) and a small reduction in the factual content of the syllabus will not be sufficient.

The lesson learnt from looking at the thirty-point assessment must be that unless teachers are given adequate training (and training colleges take this seriously) and time to carry out the task, assessment will become isolated from teaching. In the real life of classroom and laboratory, effective teaching will not take place – only assessment. The result of this will simply be a finer grading of pupils ranging from Grade A to G, with Grade G being regarded as a condemnation rather than a certification. Further, if the pedagogy remains unaffected as a consequence of insufficient funding for proper training and resourcing, GCSE will almost certainly fail to provide access to a worthwhile certificate for any of the bottom 40 per cent who at present remain without a certificate. And as the GCSE certificate will have currency value in the society (humorously depicted by the cartoon), those without will be confirmed as failures, even in a 'fairest possible' examination.

One problem with presenting an assessment scheme independently of pedagogy is illustrated by LEAG's scheme for differentiated exam papers, as featured in the GCSE Science Guide. This scheme can be adopted equally by schools which have mixed-ability classes and those which 'stream'. Surely the spirit of the GCSE's positive assessment is to enable children to have access to as much of the teaching material as possible for as long as possible, so that pupils are given the most favourable opportunity to show what they *can do*. In a 'streamed' school, candidates deemed lower-stream (the present CSE) will be taught materials to cover only Papers 1 and 2, whereas those deemed to be in the upper stream will be taught materials appropriate to Grades A to D (the present O levels). The consequent disenchantment and lack of esteem which 'stream-

GCSE? That will do nicely.

'ing' produces will remain intact within this mode of GCSE teaching.

On this vital point, once again the Science Guide fails to give clear advice:

> As the GCSE becomes established, simpler and more rational administrative arrangements should emerge. Even within the current system, some schools find it possible *to delay entry decisions* until students are well into their fifth year. Already, for example, where schools are able to offer science courses leading to Joint O-level and CSE examinations, *targets can be chosen* at a time when a good deal of information has been gathered on students' attainments. (pp. 26–7, emphasis mine)

The above jargon – about delaying entry decisions and the choice of targets – appears to suggest a non-streamed approach, yet stops short of recommending it. The document gives no further guidance on this crucial issue.

IMPLICATIONS

Should not the 'positive assessment' accompany an equally positive pedagogy? If it does not, it will simply produce a finer grading system, perhaps with one difference: the 'grades' will be seen as more 'fairly' allocated. Therefore the recipient of the lower grades, or of no grade at all, will be seen as deservedly worthless.

What we assess, how we assess it and why we assess it is inseparably linked to what we teach, how we teach and why we teach. It seems that any discussion of assessment in science (or, any subject, for that matter) must include a discussion of our notions of knowledge, learning, intelligence and ability. It also seems essential that the pupils themselves become active participants in the process of education, and – without being made to feel insecure by questioning and discussing everything under the sun – be made privy to the values and assumptions which underpin the methods of assessment they undergo.

232 Anti-Racist Science Teaching

References

(All are published in London.)

GASP Newsletter, 2 December 1984.
GASP Team, 'The Reply to the GASP Critique', *Science News* 28 (November 1984).
GCSE Science: *A Guide for Teachers*, SEC/OU, 1986.
GCSE Physics: *The National Criteria*, SEC, 1986.
B. Goacher, *Selection Post-16: The Role of Examination Results*, Schools Council Examination Bulletin 45, Methuen Educational, 1984.
Hargreaves Report, *Improving Secondary Education*, ILEA, 1984.
ILEA, *Race, Sex and Class* Booklets 1 and 2, 1983.
LEAG (London and East Anglian Group), *GSCE Examination in Physics*, Syllabus B, final approved version for 1988.

'ABILITY' LABELLING AS RACISM

LES LEVIDOW

INTRODUCTION

What is the connection between 'ability' labelling and racism? A critique could take up several different levels of connection, from the most blatant to the most subtle, as follows:
□ the use of IQ test results for making explicit claims about the supposed intellectual inferiority of black people;
□ the actual process of IQ testing and of achievement testing as oppressive; and
□ the criteria for 'achievement' embodied in formal testing but also prevalent in schooling more generally, especially in science teaching.

Critics of IQ testing have tended to focus upon explicitly hereditarian claims made about test results. Yet such claims are only the most blatant manifestations of a deeper problem. That is, the more widespread and insidious forms of labelling arise from the individualist standard of achievement that dominates schooling.

For example teachers, school psychologists, parents and even school students themselves tend to take for granted such meritocratic labels as 'clever/thick child' or 'slow/fast learner'. It can seem almost in the nature of things for people to be classified and treated according to such competitive categories (Turner, 1984). Thus ability labelling can continue its oppressive effects even without hereditarian claims or systematic IQ testing, in which the labelling is historically rooted.

The problem extends far beyond that of conscious prejudice or racial stereotypes. Teachers who believe themselves to be non-racist – and even teachers who try to overcome their own recognized racist attitudes – still unwittingly perpetuate a more subtle racism through their teaching practices. And black youth are not at all 'slow' to see through the ways teachers claim they are 'non-racist' or 'colour-blind'.

Although the traditional IQ tests are no longer used so widely as before, their models of ability have been incorporated into the prevalent methods of teaching and assessment, including the predetermined designation of failures according to the 'normal distribution curve'. In Britain today, only 60 per cent of pupils leave school with any formal qualifications – CSEs, O levels, GCSEs. Black and working-class youth are over-represented in the other 40 per cent. What does it mean to say that the schools 'fail' them? Many teachers attribute the low success rate to students' innate low ability. Others sense that the 'failure' has more to do with the way shools encourage meritocratic pride in some and lowered expectations in others, and the way our society selects its university entrants and future scientists. Many teachers are aware of these institutional pressures, yet find themselves colluding in selecting and ranking the official successes while mystifying the results as a product of 'ability' differences among students.

'Underachievers' are left wondering whether they were too stupid or lazy to succeed. Many end up feeling excluded in particular from science – not just from a scientific career, but from practising science or even holding valid opinions about science-related issues. Even though traditional IQ tests are no longer used systematically in state schools, we still need to examine how their criteria for success have continued to operate more subtly, though no less effectively, in racist ways.

The first part of this article analyses the most formal, classic version of ability labelling – the process of ability (and achievement) testing – as a power relation. Rather than criticize such testing as simply unscientific, this critique focuses on its social content and political role, which make it fundamentally similar to the predominant science in this society. The second part sketches the various notions of 'ability' contained in debates over schooling since the 1970s, towards evaluating the main proposals which claim to overcome the racist stigmas of ability labelling.

I: IQ AS CAPITALIST SCIENCE

Since the revived race-IQ debate of the 1970s, many critics of IQ testing have challenged its scientific status by denying that it measures children's true abilities, or at least those of any but white, middle-class children. For example, Bernard Coard (1971) was one of the first critics to identify cultural bias in the test situation as well as in the test questions. While certainly pointful, such criticism of bias implies that there may still be some alternative testing procedure that would be truly fair and scientific.

This article suggests that a search for such a procedure would be misguided, because the most fundamental problem is not the test as such. Rather, it is the working assumption that each person has a quantity of 'ability' which could be measured, much as scientists go about measuring the physical properties of matter. Accordingly, the testing process portrays a certain definition of ability as a natural human attribute, as a 'property' of the individual. Yet the 'intelligence' of IQ testing is constructed by the testing process itself. As I will argue, this is just one example of how science in this society tends to naturalize 'civilization as we know it', to represent aspects of capitalist society as the natural order of things.

WHAT KIND OF BIAS?

In 'ability' tests the questions often require some kinds of concrete knowledge or experience which are relatively more familiar to white, middle-class children. In that sense, they test a certain kind of achievement rather than innate ability. In reply, some test proponents assert that the nonverbal or abstract reasoning questions help to overcome possible bias, and that in any case we have no better predictor of academic success.

Of course, such predictors often act as self-fulfilling prophecies. Moreover, although it is worth criticizing IQ tests for cultural bias, it is more important to stress that achievement tests themselves incorporate a class-based model of knowledge partly derived from IQ tests. And this similarity is not simply a matter of test questions requiring certain types of concrete knowledge. Be they apparently verbal or nonverbal, apparently concrete or abstract, they all demand a lone submission to rules of abstract thought which suppress certain associations with concrete experience.

Although all children are capable of abstract thought, they prefer to approach classroom situations and test questions from combined cognitive and emotional associations with their experiences, which they tend to talk about in their own informal, everyday language. It is precisely such language and associations which the testee must suppress in order to abstract thought processes from concrete experience. Only in such a way can the test call up an interchangeable form of 'ability', indifferent to any particular social circumstance.

That 'ability', enhanced by a middle-class upbringing, involves a culturally specific form of suppression. By commanding standardized test answers which can be easily quantified, the test represents the testees' quantitative differences as the natural distribution of a universal quality. Thus class differences are made to appear rooted in natural differences in the possession of a thing – 'ability'.

It has been said, half-jokingly, that a test tests the ability to take tests. We can now see a more profound meaning in that old quip, in so far as the 'ability' entails the self-discipline to pursue formal rules of thought whose content bears no intrinsic interest for the testee. Through this form of individualized testing, the various class and cultural differences among schoolchildren are translated into quantified inequalities of presumed ability and achievement.

PEDAGOGY FOR TESTING

The official distinction between those two 'measurements' – ability and achievement – has been fundamental for legitimizing IQ testing as a predictor of academic achievement. For example, when working-class children were selected for entrance to the largely middle-class grammar schools, the scientific instrument was the 11+ exam, which often included IQ tests. We should be more precise, though: although a child's IQ score alone was a worse predictor of academic success than either the teacher's estimate or other tests, the IQ score combined with another index was a better predictor than any single one. And on occasions when a local educational authority removed the IQ test from the 11+ exam, working-class children did less well than before, and the proportion of them admitted to grammar schools dropped (Evans and Waites, 1981, p. 108).

Only in that narrow sense was the IQ test considered 'fair'. Yet its predictive power was always somewhat tautological, since academic achievement was itself measured by tests modelled after IQ tests. The use of such testing to structure schooling naturalized the resulting class-based divisions.

We should note objections raised in the 1940s to the 11+ exam – objections relevant to today's debates over selection and curriculum. Many teachers opposed the test's presumption of innate ability for selecting out the 'most able' children at age eleven and thus discouraging more advanced teaching in higher primary classes. Furthermore, according to a 1943 government White Paper, the curriculum was 'too often cramped and distorted by an overemphasis on exam subjects'. A 1949 NUT report likewise criticized both the traditional achievement tests and the modern 'standardized objective tests', whose systematic use 'exercises harmful influence on the work of the primary schools'. Aware that coaching could significantly raise the children's scores, teachers felt under pressure to emphasize a limited part of the curriculum, treated in a narrow way, by drilling students in the types of question expected to appear on the exam (Evans and Waites, 1981, p. 96).

With the transition to comprehensive education in the

1960s, the 11+ exam was gradually withdrawn from systematic usage in the state sector. Notions of a fixed, innate 'intelligence' gave way to the more flexible, admittedly cultural notion of 'ability' (CCCS, 1981, pp. 118–22). Rather than prematurely label most students as less able, comprehensives are formally egalitarian; they give students an equal opportunity to prove themselves on achievement tests. Yet those tests, historically derived from IQ tests, sustain 'ability' labelling on a competitive individualist basis. Despite the demise of the 11+, testing today continues to reproduce class divisions – in an apparently 'fairer' way.

Furthermore, classroom pedagogy is in turn geared to knowledge as defined by achievement tests, especially in secondary schools. Although some children might learn faster than others regardless of pedagogical approach, the common labelling of 'slow/fast learners' results from that particular pedagogy. And that arises ultimately from the model of ability set by the entire testing tradition. Thus the IQ test can continue to define what counts as ability, even without such tests being administered on a mass scale. Let us look more closely, then, at the model of knowledge embodied in IQ testing as a social practice. In particular, how does it convert qualitative differences into quantitative ones?

THE POWER TO MEASURE

To illustrate the social form of IQ testing, I will relate an anecdote. It arose from an atypical test question, but it helps to reveal something profound about the normal test situation.

When I was a teacher at an inner-city secondary school (in the USA), I happened to be nearby where the school psychologist was administering the Wechsler test to Joe, one of the black students. To the question, 'What is the thing to do if you find someone's purse?', Joe grinned and replied, 'Ya wan de whitey answer or de nigger answer?' I couldn't help but burst out laughing, and Joe did as well. But the psychologist remained unmoved. So Joe broke the awkward silence by politely asking the psychologist to repeat the question, so as to restore the formal test situation. The psychologist complied, and the 'measurement of intelligence' resumed according to the rules of the proper tester–testee relation.

In that incident, Joe's initial reply undermined the tester's authority, while cleverly mocking the test question. He clearly understood the racial stereotypes underlying the officially correct and incorrect answers, while making clear his suspicions about the psychologist's own views of black people. And in effect he ridiculed the test's search for a universal moral judgement about hypothetical circumstances, abstracted from concrete experience.

His momentary disruption of the test situation revealed much about its normal basis. It is a power relation between tester and testee, who must submit to the command to perform a certain social form of mental labour, alienated from the testee's experience. The test situation itself constructs its object of study – alienated mental labour, which it defines as intelligence. Without anyone having to claim that IQ scores represent the quantity of a thing, it appears that way by virtue of assigning a number to each testee and then comparing those numbers through a distribution curve.

Such 'measurement' presupposes a power relation, which takes the form of a thing – an IQ score – attributed to the testee. Thus the validity of the score depends upon the testee's subordination to the tester, as mediator of the test's logic. Through the subordination, the tester is able to command answers of a sort which lend themselves to quantification. It is only through such answers that qualitative differences among different testees can be represented as quantitative ones, as a naturally unequal distribution of a natural property called 'intelligence'.

When Joe turned the tables on the tester, he subverted the intended process of abstracting from experience. He drew upon his experience of racism in a way which mocked the universal moral code sought by the test question. As his reply defied any criterion of interchangeability; it was truly unmeasurable – however clever.

If we can imagine an analogy between that atypical test question and the abstract reasoning demanded by the questions in general, we can begin to see the problem of test bias more broadly. As well as playing upon differences in cultural experience, the test entails the willingness and capacity to submit to a certain power relation. The charge of 'bias' understates the problem by suggesting that the test distorts the child's score from its otherwise true value, in the sense of inaccurately measuring a property residing in the testee. Yet there is no such 'thing' – except as a mystification of a relation.

Consider other examples: Amerindian children who refuse to indicate that they have found the answer until all other children present are ready to do so, thus avoiding embarrassment to the others; or children who insist upon finding an answer co-operatively. In order to assign IQ scores to such children, the test process would have to suppress that cultural difference and hide it behind numbers. Of course, a society could just as well decide to assess children according to how well they help each other to learn. But that, for somewhat obvious reasons, is not the criterion that a capitalist society promotes.

THE IQ DEBATE

Not only is the test process mystifying, but so are the main issues of the IQ debate, by virtue of their terms of reference. Let us examine three of them, in turn.

1. How well does the IQ test measure intelligence?

This question takes for granted the act of supposedly measuring some physical thing residing in the testee, separated from a child's normal peer-group relations. The special power relation between tester and testee gets hidden; it gets mystified as the power of the test questions to discover the amount of this thing in each testee. Since test scores have meaning only by comparison to each other, an ordinal scale of intelligence assigns positions to each individual according to their respective amounts of this thing. Such ordinal ranking serves to naturalize society's division of labour – especially that between managers

'Ability' Labelling as Racism   241

and managed, between conception and execution – a division which appears to result from the statistical spread of IQ scores. This ranking also portrays skills as technical, neutral, separate from power relations and social values. Rather than criticize IQ tests for social bias which distorts scores, it would be more pointful to ask: What kinds of mental labour are being selected? What associations are suppressed, and what kinds of abstractions are called up? And what kind of society judges its children according to such values?

2. How much of the individual variation in IQ is due to heredity and how much to environment?

Some critics of IQ testing have used statistics to argue that the variation of IQ scores is due entirely to environmental differences. That argument begs a crucial question, beyond the statistical arguments within the 'nature versus nurture' framework. That is, why should we treat IQ as a universal natural quality? Is the very existence of this 'thing' the expression of some 'IQ genes' which then interact with some external environment? If we can see the type of power relation and mental labour embedded in IQ scores, then we can also see that the hotly debated question about 'nature versus nurture' projects that relation on to genes. In cultures where children are not socialized for competitive exams, it would be ridiculous to assume that they had genes for co-operation. Yet somehow it is considered rational to assume that we have genes for IQ, and thus to naturalize a social construct.

3. How much of the inter-generational transmission of economic success is due to the transmission of cognitive ability?

In other words, given that socioeconomic status is often somehow transmitted from parents to their offspring, how much of that transmission across generations is due to a transmission of cognitive ability?

This question was posed by Bowles and Gintis (1976), in attempting to debunk a largely unstated assumption of the IQ debate: that the 'cycle of poverty' results from low IQ. Yet they themselves make two dubious assumptions: they reduce socioeconomic status to income, and they take IQ at face value as a measurement of a purely cognitive or technical ability as distinct from social traits, which they claim influence economic success. In the course of their statistical correlations, they miss out a more fundamental connection: between the formal, 'high-status' language used to command authority over others' labour, and the types of mental labour called up by IQ testing. Both those have social aspects connected to our competitive individualist society, but such an inner connection gets lost in the statistical treatment.

CHANGING ASSESSMENT

Those three issues from the IQ debate are intended to illustrate how not only IQ testing but the predominant IQ debate itself mystifies power relations by portraying them as things or as products of things. It is not a question of whether IQ testing and its proponents' arguments are truly scientific, as if science were separable from politics. Rather, it is a matter of what kind of power it imposes, yet hides within technical terms. There would be little point in criticizing IQ testing without also challenging the power relations which make it – like most of capitalist science – 'true' and efficacious. And those relations are to be found institutionalized throughout the school system as a whole.

For example, the 'underachievement' debate involves quandaries similar to those of the IQ debate. Recalling my earlier anecdote about the recalcitrant testee, we should consider whether 'underachievement' similarly involves not simply an exclusion, but also a student resentment and rejection of the terms that the school system sets for success, in particular the way assessment divides up students. Compounding that problem is the way teachers, faced with 'discipline' problems, often try to use individual assessment as a student motivation to achieve (Scarth, 1984). This narrow definition of achievement may help to motivate some, but it certainly deters others. It substitutes impersonal formalities for real engagement with students' curiosity and camaraderie, which they learn to leave at the school gates.

What would the alternative look like? The capitalist formulae for achievement can have no simple anti-capitalist counterpart, reducible to state policy or administrative rules. Collective forms of achievement can arise only from a process of experimentation and discovery (Smith *et al.*, 1987; Baran, 1986). Some possible guidelines could be:
☐ collective self-assessment by the group of itself and the teacher;
☐ encouragement for students to share the responsibility and authority for what they learn and how, especially what questions are asked; and
☐ clarity about the political role of public examinations, regardless of who in the class ends up taking them.

The uncertainties around such a project may seem risky, and will certainly create new problems. But those problems are the ones worth solving, towards the attempt to create a society free of managerial chains of command, fixed mental/manual divisions of labour, and so on. If we do not make the attempt, we are certain to remain entrapped within the vicious circle of 'underachievement' and 'discipline' problems.

The challenge is to break the predominant type of connection between teaching, assessment and motivation and to reconnect them anew – in ways which acknowledge differences and even tensions within a group-learning process, yet without treating those as quantitative or ordinal differences. Teachers alone lack the power to relieve students of the pressures of the public examination system, which serves the higher education system and the job market. But teachers can help to clarify the real purposes of that system, while providing a different basis for students' sense of self-esteem and achievement.

II: DEBATES ON SCHOOLING

Part I showed how IQ testing mystifies a certain power relation, a certain type of mental labour, as measurement of an innate thing called 'intelligence'. It showed how that model of knowledge has become embodied in 'ability' labelling and related teaching methods, geared to the capitalist division of labour.

Part II will look at how certain notions of ability have arisen in public debates on schooling since the 1970s: from the revived race-IQ debate and ESN schools, to the 'Great Debate' and 'assessment of performance', to black 'underachievement', to multicultural education and finally to vocational training. The purpose is to evaluate how these debates have either concealed or revealed the inner connection between racist effects and schooling for capitalism.

RACE-IQ DEBATE: USA

After the Second World War, it seemed that the scientific racism of the early twentieth century had been for ever discredited by its more recent association with the Nazis' programme of slave labour and genocide. However, the whole question of a genetic basis for 'racial inferiority' again came to prominence in 1969 with the publication of a provocative article by the US psychologist Arthur Jensen. He suggested that the apparent failure of compensatory education programmes in the USA lay in the innate inferiority of the children involved, mainly black ones. As Jensen put it, using the old (pre-'black power') terminology,

> on the average, Negroes test almost 1 standard deviation (15 IQ points) below the average of the white population in IQ . . . Also, as a group, Negroes perform somewhat more poorly on the subtests which tap abstract abilities . . . the discrepancy in their average performance cannot be completely or directly attributed to discrimination or inequalities in education. It seems not unreasonable . . . that genetic factors may play a part in this picture. (pp. 81–2)

To back up his argument, Jensen cited previously published papers on IQ differences and twin studies comparing fraternal and identical twins, all of which supposedly demonstrated that 80 per cent of the variation in IQ was due to heredity. His main source of data was the British psychologist Cyril Burt, who did not concern himself with racial differences but who did use heritability arguments to justify a school system which overtly perpetuated class divisions. Now Jensen was using heritability arguments to remove personal blame from 'less intelligent' Blacks for their failure at school, while attacking those misguided reformers who irresponsibly encouraged unrealistic aspirations among such unfortunate people.

One of the main rebuttals to Jensen came from Princeton psychologist Leon Kamin (1974). He documented how the early proponents of IQ testing had both conceived the tests and used the results for reactionary purposes, especially as 'scientific evidence' to justify sterilization measures and immigration restrictions. Also – scrutinizing the major evidence wich purported to demonstrate the heritability of IQ – he argued that it proved no such thing. Furthermore, he noted that the data from Burt's twin studies seemed statistically too 'perfect' to have been for real.

What made Jensen's claim so controversial was not simply his scientific status but the policy debates then raging over compensatory education. Jensen's article was widely cited by conservatives arguing for cutbacks in programmes such as Project Head Start, intended to give black children a more equal chance in 'the race of life'. The cutbacks formed part of the Nixon administration's policy of 'benign neglect', funding black entrepreneurs rather than the social welfare programmes of the earlier Democratic administrations.

Supporters of compensatory education responded to Jensen mainly by explaining the apparent racial differences by environmental causes – poverty, family structure, racial discrimination, underfunded schools, and so on. Some went further and challenged the bias of the IQ tests themselves, or argued that IQ scores should be irrelevant to funding decisions anyway (Bowles and Gintis, 1976).

But deeper questions needed asking. What was compensatory education compensating for? Simply for disadvantages that certain children bring to the school? Or for the way the schools themselves disadvantage them? And how could compensatory education best avoid reproducing those features of schooling? These issues became marginalized amidst statistical arguments within the 'nature/nurture' debate, concerning mainly the genetic versus environmental causes of IQ differences.

RACE-IQ DEBATE: BRITAIN

In Britain, Jensen's views were popularized by Hans Eysenck, who provoked many critical responses (e.g. Rose, 1973). But here the race-IQ debate remained marginal to specifically British developments, such as controversies over ESN schools and streaming more generally.

For example the Black Papers, concerned with class more than race, challenged the move towards comprehensive schools. Their authors affirmed the general intellectual inferiority of the working class and its genetic basis; they also decried the loss of the grammar schools and thus the loss of opportunity for the most 'clever' children of the working class to rise out of it. Attacking progressive education as responsible for 'falling standards', these authors urged greater centralized control of schools in order to shift their curriculum towards traditional basic skills and to intensify both streaming and traditional examinations (CCCS, 1981, pp. 200–7).

At the same time, a different kind of protest was coming from many black parents over ESN schools, for children labelled 'educationally subnormal'. A 1968 ILEA report had recognized the disproportionate numbers of black students being relegated to such schools. An ESN teacher, Bernard Coard (1971), went on to take the most critical approach of all. He attacked a key to ESN labelling, IQ tests, as designed for white, middle-class children. He saw a cultural bias both in the test questions and in the social setting of the formal test situation. From his own experience, he observed that West Indian children performed tasks much better if they thought they were merely playing a game or helping teachers to develop teaching materials than if they knew they were taking an IQ test. Black children being assessed for ESN were made to feel rejected through racial discrimination, regardless of the ultimate assessment; furthermore, those who refused to co-operate with such a procuere were often labelled mentally retarded.

It is worth noting here that – according to later research by Sally Tomlinson (1981) – in practice low IQ scores were a minor reason given by those professionals allocating children to ESN schools. The primary basis seemed to be various beliefs about what an ESN child is. Such beliefs often coincided with teachers' images of West Indian students as 'boisterous, disruptive and aggressive'. Some of the decision-makers would label a child as ESN because s/he 'cannot communicate adequately' – which, of course, referred to communication with authority figures, not with peer-group members. Analysing the ESN category as 'a form of social control for a potentially troublesome section of the population', she suggested that its function was 'to permit the relatively smooth development of the "normal" education system'. That is, it was to justify removing certain children whose lively behaviour might be imitated by the more disciplined ones – class and cultural differences which IQ science reduces to quantitative ones.

By the mid-1970s there were growing doubts about the classic data of Cyril Burt, which hereditarians had been using to demonstrate their theories. Finally, Kamin's suspicions were confirmed in 1976 by a *Sunday Times* journalist, Oliver Gillie, who claimed that some of Burt's alleged 'co-workers' were fictitious, and perhaps their alleged data was as well. Eventually, even Jensen had to admit that they were worthless. As early as 1974 Jensen had already been musing, 'It is almost as if Burt regarded the actual data as merely an incidental backdrop for the illustration of the theoretical issues in quantitative genetics . . .' (p. 25). Indeed, it turned out that this 'incidental backdrop' was itself faked, by the main proponent of the heritability theory that the data supposedly proved.

However, it was one thing to discredit Burt's data, and quite another to refute his theory. The way remained open for others to repeat Burt's alleged studies for similar ends, starting from the same 'nature/nurture' framework. Two minor hereditarians soon published new calculations based on other data supposedly proving that 'intelligence' is 80 per cent heritable. Reporting this great scientific breakthrough, a science journalist typified the prevailing mentality when he observed, 'In today's more objective climate, controversy is more likely to be scientific than political.' It was not accepted that science itself was fundamentally political.

ASSESSMENT OF PERFORMANCE

Meanwhile, increased pressures for student assessment were arising from a different source. In 1976 the Callaghan government initiated the 'Great Debate' on schooling. Attributing high youth unemployment to their supposedly deficient skills, the government proposed reorientating the curriculum more closely to 'the needs of industry'. This meant diverting attention away from a single ordinal scale of ability, towards 'basic skills' such as numeracy and literacy. The government began to shift responsibility and funding away from the schools, and towards the Manpower Services Commission, for shaping the new kinds of labour power needed by industry. Tory leaders rightly boasted that this move conceded to the ground that they had been setting out, through such publications as the Black Papers.

But what were 'the needs of industry'? It turned out that employers, although often unclear about this, were complaining less about school leavers' technical deficiencies than about work habits, such as their reluctance to show an enthusiastic commitment and obedient attitude towards menial, low-paid jobs (Frith, 1978). So the government was overtly charging both the schools and the MSC (e.g. YOPS) with the task of inculcating the capitalist work discipline that would make youths more 'employable'. (For a broader critique of the MSC, see Bates *et al.*, 1984; Finn, 1987).

Complementary to the proposed 'reskilling' of the nation's youth, in 1975 the DES had already set up the Assessment of Performance Unit (APU) 'to promote the development of methods of assessing and monitoring the achievement of children at school, and to seek to identify the incidence of underachievement'. It was nominally intended to investigate how underachievement might relate to educational disadvantage and resource allocation. But that concern was eclipsed by its real priority: to monitor achievement and establish national standards, in response to ongoing party political debates over 'falling standards' and 'freedom of choice'. In particular, this meant identifying schools which 'consistently perform poorly', as an aspect of local education authorities' accountability for educational standards. Hypothetically, this might have meant giving extra help to under-resourced schools; in the context of overall budget cuts, it meant judging schools for their cost-effectiveness, like any business investment.

The APU's obsession with such statistical comparisons presupposed that the assessment methods had the same meaning for all children, regardless of cultural differences. It precluded any wider discussion of what causes 'underachievement', much less of how to measure what counts as achievement. Given the way journalists treated the APU results as proof of girls' intellectual inferiority, rather than question predominant teaching or assessment methods, it is just as well that the APU failed to get the co-operation it requested to investigate 'racial' differences (Hextall, 1984).

The APU denied that it was a step towards establishing a national core curriculum and thus infringing teachers' autonomy. Yet its assessment methods tended to undermine any teaching methods which deviated from treating 'ability' as fragmented, technically definable skills, measurable through individual performance on demand. For example, for the assessment of writing skills, the tasks tested included the child's 'objective description or account of a process', as well as 'expressing his [sic] feelings about it'. So testees were expected to find a supposedly neutral observation language, devoid of feelings or value judgements – as if there were such a thing. They were expected to legitimize the ideology of 'objectivity', to split their own experience into two distinct exercises – 'objective' and 'subjective' – each to be performed for the sole purpose of having it judged by a tester.

As regards assessment of 'communications skills' more generally, it attempts to separate out children's skills from their social sense of camaraderie in order to submit such skills to the formal exercise of a test. One English teacher noted the pressures being exerted upon the English curriculum for literary skills to be reduced to some abstract 'form in itself', without any intrinsic interest in the material at hand. By defining skills in that narrow way, the APU has pressurized schooling towards imposing industry's most basic skill: obeying instructions from above.

In the case of science assessment, the APU appropriated progressive education's emphasis on 'process', but did so by separating out process from content. For example, it attempted to do separate assessments of supposedly distinct stages of designing, carrying out and interpreting a classroom practical. Although these artificial separations implied no particular hierarchy of skills, by the 1980s they were becoming a basis for individualized 'graded tests' (Baran et al., 1984); they now involve mental/manual hierarchies corresponding to apparently high/low-ability students.

Some science teachers have welcomed the APU for making standards more specific or for revealing that most students of a certain age misconstrue certain concepts. In that case, we might well ask: Why wasn't such a problem obvious in the course of normal classroom teaching? And if the APU is to help as a diagnostic tool, then will the corrective teaching methods incorporate the APU's artificial separation of skills? That seems already to be happening, as the APU model is used to structure teaching in lower-stream classes not aiming for exam qualifications.

PROFILING

Since the development of the APU, a similar method of profiling students' supposed weaknesses and strengths has been applied on an individualized level in vocational training. This is done not just in MSC programmes (YOPS, YTS), but also in schools, in some cases programmes funded there by the MSC –

e.g. TVEI, the Technical and Vocational Education Initiative, aimed mainly at non-examination students. As one critic has described the transition towards an educational certificate in employability, 'The practical, the familiar and immediate, common-sense and everyday knowledge become the subject of the curriculum, displacing the analytical and cognitive, the unusual and distant, the universal' (Ransom, 1984).

Blurring boundaries between disciplines, this shift moves from a traditional academic curriculum to a more integrated one. Hypothetically, such a move could encourage students' critical powers and independence from the teacher's authority. But it can also do the opposite, as do these new pre-vocational courses, given the narrow way they define what counts as knowledge. As Basil Bernstein had warned, this means a kind of individualized profiling in which 'a new range of pupil attributes become candidates for labels. In this way, more of the student is available for control' (quoted in Ransom, p. 231).

With prospects of automation, ever-rising unemployment and underemployment, the state must control not just youths' skills and attitudes towards work, but also their aspirations. If employers need fewer academically qualified workers, then the state must not only adjust the supply but also represent the new stratifications of labour power as an obvious consequence of ability differences. As the 1980 MacFarlane Report argued, 'young people should branch out at the age of sixteen, each according to his or her abilities'. Or, as a DES official put it more bluntly, 'People must be educated once more to know their place' (quoted in Ransom). This reveals the political content of the shift from public examinations to personal profiling, as well as the connection between schooling and the MSC slave-labour schemes outside them.

At the same time, the more progressive teachers were faced with quandaries over how to avoid reproducing the traditional class-based divisions in the new comprehensive schools through streaming. While many Tories were calling for a restoration of grammar schools or more formal streaming, many teachers were supporting 'mixed-ability' classes. In principle this could have meant simply a random (e.g. alphabetical) selection procedure. In practice the term usually meant testing students for 'ability', then 'mixing' them according to the test results. Sometimes it even meant doing periodic testing to demonstrate that such classes produced more 'achievement' than streamed ones.

By conceding to such criteria, the move to 'mixed ability' has kept its proponents on the defensive and made conventional achievement criteria even more important than before. Both assessment and the curriculum itself get orientated towards proving that 'achievement' – as officially defined – does not suffer in mixed-ability classes. Imposing such priorities on these classes invariably leads to 'discipline problems', as 'underachieving' or excluded students express their frustration. Some teachers attribute the problem to the 'ability' mix as such, yet it is the individualist model of achievement that created the problem in the first place. Thus potential gains for equal opportunity have been limited by this tyranny of labelling.

'UNDERACHIEVEMENT'

By the late 1970s, the earlier debates over ESN schools had expanded to one over black 'underachievement' more generally. Black parents' groups were challenging the overtly racist practices of schools, but it was more difficult to challenge the way the schools structure achievement.

Unfortunately, that was the case for a black organization's report on Redbridge schools, Essex (BPPA, 1978). *Cause for Concern* purported to analyse the relative academic failure of West Indian youth. It strongly implicated institutional racism in hindering achievement, but not in defining what counts as achievement. Instead of asking West Indian students how they experience schooling, the Working Group considered a long list of possible causes of their low achievement. The report dismissed, among others, a genetic explanation – not because of any political objections to IQ testing, but only because

'Jensen's views had been completely refuted at an academic level'.

According to the report's conclusions, the central cause of 'underachievement' was really 'the development of a negative self-image in a hostile society', as well as other cultural factors. The Working Group considered such factors simply as the cause of an individual deficiency, but not as a source of a defiant collective self-image thrown up in defence against that hostile society.

In his critique of the Redbridge report, Farrukh Dhondy (1978) went as far as to suggest that many West Indian youths may 'underachieve' because they reject the terms the schools set for success. That is, their peer-group solidarity gives them the strength they need to resist collectively the school's discipline, which would otherwise fragment them into vulnerable individuals competing for achievement. Then again, such an interpretation could hardly be expected of a Working Group hoping that 'equality of opportunity will help to avoid black people forming a deprived subgroup which could threaten the stability of the nation'. Thus the first-ever study involving local black parents was channelled by the Community Relations Council into 'concern' for containing the revolt of black youth.

Other parents' groups have taken a more critical approach. For example, in North London the Black Parents Movement, in opposing cuts in the school budget, proposed 'independent parent power' for wider political changes in the content of schooling. Its 1979 conference document clarified the group's view of the present system:

> The B.P.M. sees *schooling as the preparation and selection of workers for the labour market.* Some for better-paid factory jobs, some for professional or middle-class jobs, others for the low-paid jobs or for unemployment. The exam system is the means by which this selection is made.

On behalf of the movement, Gus John (1986) later described the wider effects of its campaigns in the 1970s:

Grave injustices in educational practice against black kids *and* white kids were exposed. Racial bias and class bias in the determination of educational ability was found to be endemic within the education system. Here, as in so many other areas, the struggles waged by the black working-class movement were to have the effect of revealing the way the white working class had been treated for almost a century, since the Shaftesbury Act which guaranteed popular education in Britain, and were to lead to a shake-up of schooling practices for all school students, black and white.

Subsequent official reports, commissioned by the DES or ILEA, were to find one means or another of hiding those connections between race and class oppression.

THE RAMPTON REPORT

In response to widespread concern over the academic 'underachievement' of West Indian children, in 1979 the DES set up a Committee of Inquiry into the Education of Children from Ethnic Minority Groups, chaired by Anthony Rampton. The Committee gave particular attention to 'the educational needs and attainments of pupils of West Indian origin'. Its 1981 Interim Report, *West Indian Children in Our Schools*, attributed their underachievement to

> ... no single cause ... but rather a network of widely differing attitudes and expectations on the part of teachers and the education system as a whole, and on the part of West Indian parents, which lead the West Indian child to have particular difficulties and face particular hurdles in achieving his or her full potential.

Rejecting such explanations as IQ differences or linguistic differences in themselves, the report emphasized social deprivation and racial discrimination, both in the school and in society at large. It pointed to explicitly racist views among some teachers, as well as a widespread 'unintentional racism'. That is, teachers' low expectations become self-fulfilling

'Ability' Labelling as Racism 255

prophecies by demotivating West Indian pupils, who may be adversely affected by 'stereotyped, negative or patronizing views of their abilities and potential'.

According to the report, these 'negative teacher attitudes' went further than simply low academic expectations:

> ... there seemed to be a fairly widespread opinion among teachers to whom we spoke that West Indian pupils inevitably caused difficulties. These pupils were, therefore, seen either as problems to be put up with or, at best, deserving sympathy. Such negative and patronizing attitudes, focusing as they do on West Indian children as problems, cannot lead to a constructive or balanced approach to their education ...

(Note the similarity to Tomlinson's observations on ESN labelling according to social behaviour.) And these attitudes were compounded by an 'inappropriate curriculum' imposing Eurocentric approaches to cultural differences.

Although the Rampton Report feebly pleaded insufficient data to confirm or deny the disproportionate placement of West Indian children in ESN schools, it did warn against the possible cultural bias of the tests used and against racial discrimination because of cultural differences. The report recommended that pre-school facilities be extended and that schools 'reach out' more to West Indian parents.

On assessment, it proposed that all pupils' progress be monitored in order to identify differences in achievement early on, and so take remedial action. It also suggested that 'examinations have a major part to play in complementing and reflecting a multicultural approach to the curriculum in schools and the multiracial nature of today's society'.

While pointing out the racist bias of curriculum materials and exams, the Rampton Report's multiculturalist remedy limits itself to adjusting their content. Accepting that there will 'always be some children who will underachieve', it ignores the way the present form of the entire competitive exam system asserts 'the dominant culture' – that is, through teaching and assessment methods which inevitably label so many students as failures. The report, in effect, hides that connection by emphasizing 'the particular educational needs' of particular ethnic groups.

THE HARGREAVES REPORT

Another relevant study is the Hargreaves Report, *Improving Secondary Schools*, published by ILEA in 1984. It has been welcomed for the progressive measures that it proposes, such as 'group work, co-operative learning and encouraging pupils to find out for themselves'. Yet it founders on the problem of underachievement. For example, it proposes that 'pupils should be equally esteemed whatever their academic ability' – still ambiguous about labelling according to ability hierarchies.

The Hargreaves Report also acknowledges that 'public examinations are an inadequate measure of educational achievement'. However, it does not criticize the individualist definition of achievement embodied in such exams; instead it takes apparent 'ability' differences for granted and falls back upon 'deficit' explanations. It attributes underachievement to students' insufficient motivation, in turn due to their disadvantaged family background. Assuming that working-class pupils come to school with low levels of motivation, the report argues that such pupils rely on teachers and exam results for their sense of self-esteem.

For enhancing self-esteem it proposes 'graded tests', designed so that all students can feel confident of passing the easiest of these and feel encouraged to progress as far as they can. Thus the APU model of fragmented knowledge is extended to a hierarchy of abilities, as a basis for individualized profiles. (For its application to science, see our critique of GASP in this collection.)

The Hargreaves Report can make such a proposal only by ignoring the way exams deter motivation. That is, the pressure of exams tends to destroy whatever curiosity and internal motivation students may bring to their school subjects, as well as threatening to divide up students from each other. When the report acknowledges that underachieving students 'show their disaffection with school by absenteeism or other uncooperative behaviour', the authors seem not to realize that such behaviour often involves 'co-operation' among students to defend themselves from their fear of being labelled failures or from being divided up. Rather than confront that problem, the Hargreaves Report evades it through a minor tinkering with the exam system. Thus it turns out to limit the progressive potential of its proposed reforms of teaching methods.

In discussions with Hargreaves himself, some teachers have criticized the report for neglecting racism and sexism or for subsuming them under class. In reply, Hargreaves has pleaded lack of time and resources, while warning against 'utopian critiques' devoid of practical import. Yet, as we have seen, the problem lies deeper, in the report's notion of class as a family deficiency which undermotivates children to learn. By taking such a patronizing attitude towards working-class students, and by proposing yet additional types of individual assessment, its proposals cannot get very far in alleviating divisions based on racism, sexism or class.

Furthermore, the Hargreaves Report fails to take up the implications of ILEA's own previous statements on assessment. In its *Race, Class and Sex* booklet no. 1, ILEA clearly implicated a secondary school curriculum 'dominated by public examinations designed to mark pupils into fine grades on the basis of performance in timed tests'. Even if the new proposed 'graded tests' were to avoid such domination – which seems unlikely – they still would do nothing to challenge the competitive system of public examinations, which ILEA acknowledges is responsible for sustaining 'the correlation between social disadvantage and educational attainment'.

THE SWANN REPORT

Following publication of the Rampton Report, the government had replaced the Committee's chair with a presumably safer appointee, an ex-BBC chairman, Lord Swann. The resulting Swann Report, *Education for All*, was published by the DES in 1985. Extending the original Rampton survey to all 'ethnic minority' children, it defined the problem more broadly than black underachievement, as it saw many children of all kinds as 'not achieving their full potential'. It noted a widespread educational disadvantage due to socioeconomic deprivation, which was compounded by racial discrimination in the case of West Indian children. It attributed underachievement to 'deeply rooted prejudice against all kinds of ethnic minorities' on the part of Whites.

Although its criticism of schools went no further than its predecessor's (Rampton) report, the Swann Report provided extensive documentation of racism in schools. In particular, interviews with black students who made it to university reveal how West Indian students find their cultural differences converted into racist labels – such as 'rude boys' or 'truants' – or suppress those differences for the sake of academic achievement. Some examples:

> I mean – a lot of black kids get into trouble because they are truanting because often the lessons are so boring, the teachers are not giving them any stimulation . . .
>
> I personally denied my blackness, because that is how I made it in the system . . .
>
> I look back . . . at the years that I have wasted . . . spending hours with things which are totally irrelevant to me and to what I want to do with my life as a black person . . .
>
> They force white middle-class values on you, that's the only way they think you can succeed. (pp. 93–103)

Notwithstanding the report's call for teachers to recognize the special needs of West Indian pupils, these experiences suggest that schools disadvantage them because their strong cultural identity provides strength for resisting the individualist rules of the game, thus setting a dangerous example for others. In view of those experiences, it is worth looking at how the same report treats cultural bias in testing.

In a detailed appendix on the IQ controversy (pp. 126–63), it suggests that IQ tests as such are not especially biased against

West Indian children. Taking a narrow definition of bias, it argues that any test of intellectual potential presumes some kind of knowledge, and the IQ tests seem not to underestimate these children's future school achievement. While it acknowledges that the difference between IQ tests and school exams 'is a difference of degree, not of kind', it notes the lack of any alternative measure of intellectual potential. Thus it accepts IQ tests 'as good a measure of intelligence or cognitive ability as we have'.

In this way it begs the question of why intellectual potential and achievement alike should be assessed through such individualized quantitative 'measurement', and it reduces the issue of bias to a correlation between two such measurements. As regards cultural bias, it acknowledges that West Indian children do relatively better on nonverbal tests than on verbal ones, and that British-born Blacks do better than those born abroad. Yet it dissolves cultural differences into a vague notion of social disadvantage: 'Ethnic differences in IQ scores are probably largely caused by the same factors as are responsible for differences in IQ within the white population as a whole' (p. 148). Furthermore, while acknowledging the effects of racial discrimination, it also lends credence to hereditarian explanations for IQ differences! So, even though it calls for alleviating the factors which disadvantage black pupils, its narrow terms of reference undermine the report's own criticism of ESN labelling and 'ability' labelling more generally. Thus the appendix in effect absolves testing, and its model of knowledge, of responsibility for perpetuating the racist stereotypes that the overall report criticizes.

MULTICULTURALISM

For most educationalists concerned with overcoming black 'underachievement' (or racism), the proposed solutions include multicultural education. Its proponents hope that it will help ethnic minorities to fulfil more of their academic potential. For example, they say, less Eurocentric teaching materials would make the curriculum more attractive to them while overcoming the popular prejudice that currently discourages them. Does that really mean challenging racism?

Multiculturalism can be judged only by the standard of anti-racism. In the case of science, an anti-racist approach could investigate the way Western-imposed 'development' has variously promoted or suppressed certain aspects of Third World cultures for the sake of economic domination. Students could also compare ways in which their own popular cultures – English, Irish, Welsh, Caribbean, Asian, etc. – have provided sources of resistance to oppression. Thus students could earn mutual respect by learning from each other's traditions. At the same time they could better understand the dominant, racist British culture as an obstacle to that knowledge and to a potential common interest. Indeed, given the experience of the 1981 and 1985 multiracial uprisings, some authorities fear that white youths will learn too much from rebellious black youths, both in and out of school.

Yet, by default, multicultural education has tended to mean reducing racism to a problem of omission or insensitivity in the curriculum. Accordingly, the proposed solution is to add on neglected cultures in order to correct the imbalance among a plurality of otherwise 'equal' cultures. By obscuring the ways in which cultures develop in relation to class domination, such a 'formal equality' perpetuates it. Furthermore, multicultural education has opened up yet another area of professional expertise, parasitizing and deflecting demands for a truly anti-racist curriculum. In one revealing example, where a white teacher did her bit for cultural diversity, her black pupils complained:

> ... And when we made Jamaican patties, they didn't come out in the way we wanted them to. They came out like Cornish pasties ... [The teacher] told us how to do it, when we *know* how to do it. But when we did it *her* way it came out like English food. And it's supposed to be *Jamaican* patties ... – We want to make proper patties. (*TLK* 11, pp. 19–21)

That anecdote indicates a shift to a more subtle form of racism. Whites do not necessarily claim a racial or cultural superiority, only a superior knowledge of black people's cultures (Sivanandan, 1983). It is worth probing the concept of 'knowledge' contained in the racist assumption about what it means for experts to know other people's cultures, somehow divorced from class and racial oppression. No doubt such 'knowledge' would embody the same idealized dichotomy of objectivity/subjectivity that we encounter in both APU criteria and IQ testing. In other words, cultural differences – in reality expressing relations of domination and rebellion – get reduced to merely 'subjective' differences or feelings in an innocuous pluralistic context. Any attempt at creating an anti-racist curriculum will have to confront the criteria of knowledge presumed by 'multicultural' education. Like multiculturalism more generally, it reduces racism to a problem of special needs and racist attitudes, while obscuring teachers' assigned role of shaping skills and aspirations for the capitalist labour market.

'SPECIAL NEEDS' IN VOCATIONALISM

Parallel with graded assessments and multicultural education, there has arisen a new vocationalism that also imposes racist definitions of ability through a 'special needs' jargon. Initiated from the Manpower Services Commission, it has many features in common with DES initiatives.

While vocational training has traditionally involved specific occupational skills leading to specific jobs, the new vocationalism instead attempts to train youths for transferable skills, for interchangeability, for work habits and discipline in their own right. Such 'transferable skilling' complements the deskilling, reduced bargaining power and underemployment which the working class has been undergoing with the advent of automation and monetarism (P. Cohen, 1984).

These new vocationalist schemes take up disproportionate numbers of girls and black youth, fewer of whom (than white males) find jobs upon completing them. We have already seen how the 'special needs' jargon of multicultural education sometimes implies cultural deficiencies. Similarly, the 'social and life skills' component of vocational training is offered as compensatory education for supposed deficiencies – in motivation, skill or discipline – endemic to the sorts of school leavers who end up on the schemes (P. Cohen, 1984).

One such MSC scheme, the Technical and Vocational Education Initiative (TVEI), has established itself in the schools. This precedent signifies a broader tendency for so-called vocational training to redefine schooling itself. Although TVEI declares itself 'open to children of all abilities', the pressures of rising unemployment and streaming make it seem most attractive to 'non-examination' students. As the MSC director, David Young, has so explicitly put it, 'A lot of young people are not naturally academic, so they major in truancy. They leave school unprepared for the world outside.' Indeed, the supposedly unified GCSE embodies not only the previous hierarchy between CSE and O levels, but also a third divide which demarcates the even lower-ability 'vocational' end (Chitty, 1986).

In the MSC's Youth Opportunities Programme of the 1970s, the disproportionate numbers of black people taking part were seen to confirm that they were the least 'able', the least 'employable', and thus in special need of work preparation. Now that the MSC has greatly expanded the Youth Training Scheme to include craft apprenticeships for some (mostly white) youths, disproportionate numbers of black youths are channelled into the Mode B, which is far less likely to lead to a job. As a typical comment from a YTS manager: 'Mode B provision is needed for those with special needs . . . the emotionally disturbed and ethnic minorities.' Accordingly, sometimes 'creaming off' schemes exclude such people through formal selection procedures such as tests, education qualifications and motivation criteria (Pollert, 1986). Yet even without such formal procedures, the common-sense racist labelling has its devastating effects in both denying job opportunities and attributing that denial to the cultural or linguistic deficiencies of black people.

These effects of the new vocationalism result from no racist conspiracy. Rather, it is a matter of further institutionalizing an everyday racism already built into ability labelling. The new vocationalism cannot in practice overcome such racist stereotypes for as long as it carries out its intended aim of adjusting the labour supply and youths' aspirations to a declining, deskilled labour market. Only by breaking with that political mission could it possibly train young people in skills that truly empower them – be they traditional craft skills or the political skills for understanding their predicament and for defending themselves from racist and class exploitation.

BEYOND 'UNDERACHIEVEMENT'

In conclusion, existing educational proposals fail to challenge the racism of 'ability' labelling. Predominant diagnoses fall short of identifying the competitive individualist ethos responsible for such labelling. Thus their proposed anti-racist remedies cannot challenge that fundamental basis of the British education system, which needs 'ability' labelling in order to fulfil its assigned role of shaping marketable skills and life-aspirations for the capitalist labour market. Any strategy for anti-racist teaching must challenge the real basis for black 'underachievement'. More broadly, it must also challenge the way that potentially progressive notions – such as 'child-centred' pedagogy and 'special needs' – have been appropriated for enhanced state control over schooling for capitalism.

Having seen how the predominant approaches help to perpetuate racist and class oppression, teachers may be better equipped to develop practices which challenge them. Unfortunately, even sympathetic teachers may find themselves carrying out the process of ability labelling because of the pressures of the school system. In particular, teachers sometimes justify conformity by citing complaints that any deviation would result in sacrificing pupils' future interests for the sake of teachers' political ideals. However, as this article has argued, it is the present system that sacrifices the real potential of most pupils. As the Black Parents Movement has said, 'Chasing exam passes will not bring a solution to the schooling problems faced by the majority of black working-class parents and their children' (*Battlefront*, 1986).

BIBLIOGRAPHICAL NOTE

Rather than make an ambitious attempt to analyse the origins of mental testing and its historical use in British education, this article offers a guide to some relevant publications (see Bibliography).

Gould's book takes the trouble to scrutinize in detail the calculations, statistical models and social assumptions of those who founded IQ science. Nik Rose's article suggests how the reactionary social mission that informed their methods actually constituted the formation of psychology more generally. Evans and Waites's book also criticizes those methods, as well as their later role in selection for schooling. Sutherland's book, while hardly critical, carefully documents the pre-war use of the 11+ exam and debates around it. Brian Simon, a central figure in education debates, has written several articles and books against the 11+ exam.

The CCCS Education Group, while giving little attention to testing methods, provides insightful political analysis of post-war changes in educational provision in selection. Cohen and Manion's book (Chapter 3) surveys current theories of underachievement. Broadfoot's edited collection contains several relevant essays on assessment today.

Bibliography

(All books are published in London unless otherwise noted.)

Grazyna Baran, 'Teaching Girls Science', in Maureen McNeil, ed., *Gender and Expertise/Radical Science* 19, Free Association Books, 1986, pp. 87–102.

Grazyna Baran *et al.*, 'GASP! A Critique', *Science News*, 28 (November 1984), 22–3.

Inge Bates *et al.*, *Schooling for the Dole? The New Vocationalism*, Macmillan, 1984.

Battlefront: Paper of the Black Parents Movement, no. 1, May 1986.

Caroline Benn and John Fairley, eds, *Challenging the MSC On Jobs, Education and Training*, Pluto, 1986.

Black People's Progressive Association and Redbridge Community Relations Council, *Cause for Concern: West Indian Pupils in Redbridge*, 1978.

Samuel Bowles and Herbert Gintis, *Schooling in Capitalist America*, Routledge & Kegan Paul, 1976, especially Chapter 4, 'Inequality and the Meritocracy'.

Patricia Broadfoot, ed., *Selection, Certification and Control: Social Issues in Educational Assessment*, Falmer Press, 1984.

Clyde Chitty, 'TVEI: The MSC's Trojan Horse', in Benn and Fairley, pp. 76–98.

CCCS Education Group, *Unpopular Education: Schooling and Social Democracy in England Since 1944*, Hutchinson, 1981.

Bernard Coard, *How the West Indian Child is Made Educationally Sub-Normal by the British School System*, New Beacon Books, 1971.

Louis Cohen and Lawrence Manion, *Multicultural Classrooms: Perspectives for Teachers*, Croom Helm, 1983, especially Chapter 3.

Philip Cohen, 'Against the New Vocationalism', in Bates *et al.*, pp. 104–69.

Farrukh Dhondy, 'Resisting Success: The Blacks of Redbridge', *Teachers Action* 10 (July 1978), 14–15.

Farrukh Dhondy, *The Black Explosion in British Schools*, Race Today Publications, 1982.

Brian Evans and Bernard Waites, *IQ and Mental Testing: An Unnatural Science and its Social History*, Macmillan, 1981.

Raymond Fancher, *The Intelligence Men: Makers of the IQ Controversy*, Norton, 1985.

Dan Finn, *Training Without Jobs: New Deals and Broken Promises*, Macmillan, 1987.

Simon Frith, 'Youth Unemployment and Education', *Socialist Teacher* 5 (Summer 1978), 6–7.

Howard Gardener, *Frames of Mind: The Theory of Multiple Intelligence*, Paladin, 1985.

GASP Team, 'The Reply to the GASP Critique', *Science News* 29 (March 1985), 29–30.

Dawn Gill, 'Anti-Racist Education', *Contemporary Issues in Geography and Education*, 1, 1 (Autumn 1983), 6.

Oliver Gillie, 'Sir Cyril Burt and the Great IQ Fraud', *New Statesman*, 24 November 1978, 688–94.

Stephen Jay Gould, *The Mismeasure of Man*, Norton, 1981.
Hargreaves Report, *Improving Secondary Schools*, ILEA, 1984.
Ian Hextall, 'Rendering Accounts: A Critical Analysis of the APU', in Broadfoot, pp. 245–62.
ILEA, *Race, Sex and Class* booklets: No. 1, Achievement in Schools, No. 2, Multi-Ethnic Education in Schools, 1983.
Arthur Jensen, 'How Much Can We Boost IQ and Educational Achievement?', *Harvard Educational Review* 39 (1969), 1–123.
Arthur Jensen, 'Kinship Correlations Reported by Sir Cyril Burt', *Behavior Genetics* 4 (1974), 1–28.
Gus John, 'The Black Working Class Movement in Education and Schooling', Black Parents Movement, 1986 (c/o 76 Stroud Green Road, London N4).
Leon Kamin, *The Science and Politics of IQ* (1974), Penguin, 1977.
James Lawler, *IQ, Heritability and Racism: A Marxist Critique of Jensenism*, Lawrence & Wishart, 1978.
Les Levidow, 'A Marxist Critique of the IQ Debate', *Radical Science Journal* 6/7 (1978), 13–72.
Les Levidow, 'IQ as Ideological Reality', in Levidow, ed., *Radical Science Essays*, Free Association Books, 1986, pp. 198–213.
Anna Pollert, 'The MSC and Ethnic Minorities', in Benn and Fairley, pp. 177–200.
A. Rampton, Interim Report, *West Indian Children in Our Schools*, HMSO, 1981.
Stuart Ransom, 'Towards a Tertiary Tripartism: New Codes of Social Control and the 17+', in Broadfoot, pp. 221–44.
Nikolas Rose, 'The Psychological Complex: Mental Measurement and Social Administration', *Ideology & Consciousness* 5 (Spring 1979), 5–70.
Steven Rose *et al.*, 'Science, Racism and Ideology', in *Socialist Register 1973*, Merlin, 1973, pp. 235–60.
John Scarth, 'Teachers' Attitudes to Examining', in Broadfoot, pp. 83–102.
Brian Simon, *Intelligence, Psychology and Education: A Marxist Critique*, Lawrence & Wishart, 1971.
Brian Simon, *Does Education Matter?*, Lawrence & Wishart, 1985.
Brian Simon and W. Taylor, eds, *Education in the 80s*, Batsford, 1981.
A. Sivanandan, 'Challenging Racism: Strategies for the '80s', *Race & Class*, XXV, 2 (Autumn 1983), 1–12.

A. Sivanandan, 'RAT and the Degradation of Black Struggle', *Race & Class*, XXVI, 4 (Spring 1985), 1–34.
Henry Smith *et al.*, 'Science Curriculum Innovation at Holland Park School', in this collection.
Stephen Strickland, 'Can Slum Children Learn?', in Carl Senna, ed., *The Fallacy of IQ*, NY, Third Press, 1973, pp. 150–9.
Gillian Sutherland, *Ability, Merit and Measurement: Mental Testing and English Education, 1880–1940*, Oxford, Clarendon, 1984.
Swann Report, *Education for All*, HMSO, 1985.
Monica Taylor, *Caught Between: A Review of Research into the Education of Pupils of West Indian Origin*, Windsor, NFER/Nelson, 1981, especially pp. 49–59.
Teaching London Kids 11 (1978), special issue on racism.
Sally Tomlinson, *Educational Subnormality: A Study in Decision-Making*, Routledge & Kegan Paul, 1981.
Glenn Turner, 'Assessment in the Comprehensive School: What Criteria Count?', in Broadfoot, pp. 67–82.

Bhopal

INTRODUCTION

The Bhopal disaster provides a stark example of racist science and technology. There the Union Carbide Corporation (UCC) took great risks with the lives of Indian people for the sake of profitably producing pesticides which in turn are used to make Third World agriculture more dependent upon agribusiness. Furthermore, some apologists for UCC have attributed the disaster to failings by the Indian government or the Bhopal workers.

In this section Barbara Dinham firmly puts the responsibility on to the firm which designed, built and operated the plant. Then, in greater detail, Barry Castleman documents the double standards in design between UCC's Bhopal and Institute (West Virginia) plants. Lastly, Tara Jones suggests that – despite those differences in the plants – the August 1985 leak at the Institute plant illustrates certain similarities arising from the dual drive to maximize profit and subordinate the workforce to managerial authority.

Particularly revealing is how, in both cases, the 'safety' systems hid the incipient problem and then even aggravated it. Indeed, a similar dynamic operated in the Three Mile Island nuclear disaster. In this way we can see the racist character of the Bhopal plant as a special case of the generally oppressive design and operation of chemical process industries as a whole.

MASS DEATH AT BHOPAL: WHOSE RESPONSIBILITY?

BARBARA DINHAM

The gas leak at Union Carbide's chemical plant in Bhopal claimed over 2,500 lives within a week. The Union Carbide India Ltd Employees' Union believe that the total could be thousands higher. It is said that the shroud-makers' union in Bhopal issued some 25,000 shrouds during the week of the disaster. People are still dying from the effects of methyl isocyanate (MIC) poisoning. Estimates suggest that over 10,000 people have died and over 200,000 have suffered health damage.

Press reports in the West have done much to play down the responsibility of Union Carbide for the horror of Bhopal. Initial press coverage in the UK assured us that the gas was not highly toxic, that the effects were temporary, and that Union Carbide did all in its power to prevent the possibility of 'accidents'. However, the company has a long record of negligence at the plant, and the union was warning of disaster three years before the December 1984 gas leak.

NONE OF THE SAFETY DEVICES WERE OPERATING

Management has claimed that standards at the Bhopal plant were the same as those in the United States. Yet this is quite untrue; the plant differed from other UC plants in very significant ways. The US plant has a computerized pressure/temperature system which would have made it virtually impossible for the temperature rise and increased pressure to go undetected. At UC's plant in Béziers in southern France, the automatic system is sensitive enough to detect the presence of MIC in the air at 0.3 parts per million. Leaks are backed up by an elaborate sprinkler system capable of delivering 80,000 gallons an hour. In a Bayer plant in Dormagen, West Germany, a leaking tank can be smothered in foam within eight minutes of the alarm sounding.

At Bhopal there was no plant-wide warning system; no means of rapidly cooling the tanks; and none of the safety devices was operating. Although the total picture is incomplete, the likely explanation of the leak is that a chain of events set off a chemical reaction. A pipe leading out of the MIC tank was being routinely washed. To prevent water leaking back into the tank, slip blinds are normally inserted by hand, i.e. there is no automatic system of controls and checks to prevent water leaking backwards through the valves into the tank containing liquid MIC. Water reacts with both MIC and phosgene (another toxic chemical also stored at Bhopal).

A relatively small amount of water would have started a chemical chain reaction, giving off heat. Since the liquid MIC stored in the tank has a low boiling point, the reaction would have led to vaporization, resulting in an enormous increase of pressure, and eventually to the bursting of the rupture disc and outpouring of MIC gas.

MIC should be stored at a temperature of 5°C, but the refrigeration jacket surrounding the tank had not been functioning for at least six months, leaving MIC at a higher-than-safe temperature. Should MIC escape from the tank, it is supposed to be neutralized in a caustic soda scrubber. But this scrubber was not operating. Any toxic gases which in turn escaped from the scrubber are supposed to go to the flare tower, where a pilot flame would burn the gases before their release into the atmosphere. The pilot flame was off and the pipeline to the flare tower was disconnected.

MIC should not have been stored unless at least two of the fail-safe devices were in operation. Yet even if the refrigeration unit, scrubber and flare tower were operating, the design of the plant is such that 40 per cent of the stored MIC would still have escaped. The capacities were not designed to respond to a total emergency. Water sprinklers which were designed to neutralize any leakage of toxic gas were not even capable of reaching the height at which the leak occurred. There were no gas detectors on site: operators have been told to 'use their noses' as detectors.

PROFIT VS SAFETY

Far from being a careless oversight and an unfortunate accident, the lack of operating safety measures was a consequence

of a cost-reduction drive. Only one corporate headquarters safety audit is known to have been done over the seven years of UC India operations. The last check, in 1982, exposed widespread hazardous conditions, including:
☐ lack of fixed water hose in MIC handling area;
☐ non-operational pressure gauges on the phosgene storage tank and other tanks;
☐ deadened flange points where accumulations of material might cause hazards;
☐ high personnel turnover and untrained people doing hazardous tasks;
☐ workers trained by rote memorization in hazardous tasks;
☐ lack of automatic controls to prevent overfilling of the MIC storage tank by manual methods.

The union observed, however, that the cost-cutting drive had increased since 1982–3; that is, it dated from the period when headquarters could have insisted on increased spending on health and safety. There have been no further safety audits since these hazards were revealed.

UC's safety record – already abysmal before the audit – continued to plummet. In October 1982 gas was released into the surrounding area. In 1983 there were two further leaks, and in January 1984 a factory worker died of what was euphemistically recorded as 'a chemical allergy'. The union had pleaded for safety provisions. They had written to Union Carbide management, complained to the police, the MP for Agriculture, at least two Ministers, the governor of Madhya Pradesh, the Chief Inspector of Factories, and the Home Minister of the government. As recently as January 1984 the union sent representatives to the Chief Minister. After an accident in 1982, the union had printed over 6,000 posters warning of the impending danger and posted them all over Bhopal. In October 1982 workers launched a hunger strike to draw attention to the plant's dangers. A tripartite committee was then set up to go into their demands, but officials investigating them praised UC management and ascribed the union's fear of disaster to 'imagination'.

QUESTIONING STRATEGY

What is somehow more amazing than UC's slack management of the plant, negligence, and total concern with profitability, is the absence of any knowledge about the nature of the product. MIC is described in a UC handout as having 'excellent warning properties'. Although it can be detected in the atmosphere at low concentrations, it also causes injuries at very low concentrations. By the time you are aware of it you are probably already injured.

By 2 a.m. on 3 December people were pouring into the local hospital with intense pain and burning in the eyes, coughing and difficulty in breathing, local burns on the skin, nausea and vomiting. The casualty rate was very high, and the majority of those who died were children.

At this point, clear guidance from UC on the nature of the poisoning chemical could have made a difference. UC's chief medical officer maintained until 9 a.m. on 3 December that 'MIC is an acute irritant, but certainly not lethal.' Jackson B. Browning, UC's director of health, safety and environmental affairs, said that it was 'nothing more than a potent tear gas'. Yet in fact MIC is rated five times as lethal as phosgene, a fatal mustard gas used in World War 1. The threshold limit value (TLV) or maximum permissible exposure limit to MIC is one fiftieth of one part per million, whereas for phosgene the TLV is one tenth ppm. Both are many hundred times more toxic than something as lethal as hydrogen cyanide.

There was a constant attempt to play down the lethal nature of MIC and to stress that the effect was temporary. This added intense confusion, and claimed many lives. There is absolutely no doubt that people in Bhopal have been exposed to concentrations of MIC many hundreds of times more than the threshold limits. There is a very real possibility of delayed effects, even among those who have not shown acute symptoms of poisoning.

MIC destroys the lung tissue, leading to pulmonary oedema – an accumulation of fluids – which in effect causes victims to drown. The body is also deprived of oxygen. People described how their body temperatures rose to unbearable levels, and they felt on fire. Many of those found dead had torn off their clothes in an attempt to find relief.

The company – while trying to play down the effects of MIC – could not deny the effects on lungs and eyes, but minimized any other effects. One US doctor denied that MIC is absorbed into the bloodstream, and the medical director of UC said that the chemical could not reach the liver or uterus. These statements are unsubstantiated and not based on observation or tests.

DEATH TOLL STILL RISING

While the disaster is fading from the headlines, in Bhopal the effects are far from temporary. There is evidence that MIC does directly affect the bloodstream and that either it or its breakdown products, like monomethylamine, can cause considerable chemical damage. An outbreak of jaundice among some gas victims suggests that serious damage to the liver could have resulted. Neurological symptoms such as nervousness, irritability, depression and loss of memory are common.

Many pregnant women have had spontaneous abortions and others complain of a persistent white discharge. There is no reason to suppose that a chemical which can penetrate the skin cannot enter the uterus, where its effect on the foetus would be even more devastating than its effect on children.

The union believes that the death toll has now passed 8,000 and is still rising. Yet there is still no attempt to investigate the long-term effects of MIC poisoning and ways of treating it. Thousands are still in relief camps needing daily treatment and are physically extremely debilitated. Despite this, committees have been formed and protest camps and marches launched demanding compensation, homes for orphaned children and proper medical treatment.

This is the worst industrial 'accident' in history. Yet the company adopts a most sanguine attitude. One of its corporate directors was recently quoted in the *Wall Street Journal* saying, 'While the Bhopal tragedy is without precedent . . . considering insurance and other resources available, the financial structure of the company isn't threatened in any way.'

Agribusiness is big business. Those who profit cannot afford to allow questions to be raised about the industry, the lethal nature of the products, or the possible effects of the increasing dependence of the world's agricultural production on chemicals. Business will not want to see Union Carbide sink as a result of the Bhopal disaster.

BHOPAL: CAMPAIGNS

The disaster at Bhopal is not an isolated incident, though its size and magnitude are shocking. In Brazil 2,500 pesticide and agricultural workers died of pesticide poisoning in 1984 alone. A conservative estimate places deaths from pesticide poisoning in Third World countries at 10,000 a year.

The issue of hazardous industrial production and its movement around the world must be taken up by the labour movement as a whole. This should include, as it already does in the USA, trade unionists getting together with environmentalists and consumers. This kind of broader campaign has started to happen to some extent in Britain in campaigns over asbestos and pesticides – where alliances of trade unionists, tenants' organizations, consumers and environmentalists are proving very effective.

Groups are also organizing to increase the international sharing of information. In a January 1985 meeting in Milan, a conference of pesticide and agricultural workers from different countries around the world demanded that companies should be compelled to give full information about the exact locations of plants producing, or using, highly toxic materials.

While various company and government enquiries get under way, a number of grass-roots groups are now organizing practical action specifically in response to the Bhopal disaster. In Bhopal itself local workers and residents who survived the poisoning have set up the Poisonous Gas Incident Struggle Forum. The Forum is receiving help from two Delhi-based research groups, the Society for Participatory Research *(PRIA)* and the Centre for Science and the Environment. These groups are involved in three areas of work:

1. A survey in Bhopal, where it is hoped that as many as 1,000 people will assess the real damage caused by the MIC gas leak, and what really happened. There are local estimates that many more people were killed by the gas than official figures state, because many never made it to the hospital or the official body count.

2. PRIA is planning to prepare and publish a report on the Bhopal incident as a step in building a national movement around the issue. It will appear in English and Hindi, and will be made available to unions and rural action groups across India to promote discussions and meetings.

Demonstration outside Union Carbide's Bhopal plant

3. One of the moves among such groups in India is to press for what they are calling a 'Freedom of Information Act' (similar to 'right to know laws' in the USA). In India it can be extremely difficult to get information from either government or corporate sources.

In the United States a labour resource centre called The Highlander Center based in Tennessee, in collaboration with these Indian groups, is sending technical information needed on the chemical processes involved in the Bhopal incident.

Its members are joining PRIA in preparing a report on international pesticide production, chemical use in agriculture, the role of multinational corporations (and specifically Union Carbide's operations worldwide). The Centre is appealing for information from workers in all countries where Union Carbide operates about the company's record on industrial relations, pollution, relations with the community, etc.

The centre has already produced a videotape which looks at the response by people in the USA who live by Union Carbide plants. John Gaventa of the Highlander Centre told ILR,

> We have found videotapes to be a very effective form of communication. Since the news from Bhopal, we have been making a video with people in communities in the US also affected by Union Carbide operations or the MIC chemical which leaked at Bhopal. The tape shows their responses to Bhopal, their fears that something like it could happen in their own communities, and the struggles they are involved in to make sure that nothing like Bhopal will happen again, anywhere. The tape is also being sent to India for educational use there.

Much of the information for this article was supplied by the Union Research Group in Bombay.

This article was first published in *International Labour Reports* No. 8, 1985.
This journal, published six times per year, contains up-to-date information, on trade, aid, and multinational companies.
Individual subscriptions £10, institutions £20 from International Labour Reports, 300 Oxford Road, Manchester, M13 9NS.

UNION CARBIDE'S DOUBLE STANDARDS

BARRY CASTLEMAN

This report criticizes the safety standards of Union Carbide's Bhopal plant, as contrasted with its counterpart US plant, both of which process the toxic chemical methyl isocyanate (abbreviated here as MI).

The Bhopal plant lacked any computerized pressure and temperature sensing system, as used for several years at the US plant. The Bhopal plant had widespread hazardous conditions, seen in the 1982 audit by corporate headquarters, including:
☐ lack of fixed water hose in MI handling area;
☐ non-operational pressure gauges on phosgene storage tank and other tanks;
☐ dead-end flange points where accumulations of material might cause hazards;
☐ high personnel turnover and consequent problem of untrained people doing hazardous tasks;
☐ workers trained by rote memorization in hazardous tasks;
☐ lack of automatic controls to prevent overfilling of the MI storage tank by manual methods.
It lacked anything approaching an emergency plan for evacuating and otherwise protecting the community, as exists in the USA.

Its record of plant accidents was much worse than in the US MI plant:
USA: no deaths in 17 years of MI use;
   in 1978 there was a phosgene spill – 1 person treated at the dispensary and others sent for chest X-rays;

also in 1978 MI was released when a line broke while loading tank cars – 1 man hospitalized for 2 days, 13 treated at the plant dispensary.

India: 1977 – plant opened
 1978 (24 December) – huge fire in naphtha storage area
 1981 (26 December) – phosgene leak, 1 man killed
 1982 (January) – phosgene release left 24 people severely ill, including plant neighbours
 1982 (5 October) – flange broke and gas release caused mini-riot in shanty towns around the plant (mixed MI, HCl and chloroform)
 1983 – two 'minor leaks'
 1984 (January) – a factory worker died of 'a chemical allergy after working at the plant'.

Its engineering control equipment had not been working for a long time before the 3 December 1984 release of toxic gas:
☐ MI refrigeration unit had not been in operation since 1982;
☐ caustic scrubber and flare had been out of service for over one month, in violation of Indian factory rules.

In addition to the failure of the above engineering systems, the alarm which was supposed to warn the community did not sound until half an hour to two hours after the release of gas. This alarm had gone off accidentally in the past, and the citizens in the neighbourhood were clearly not apprised of its significance.

Underqualified people were running the plant at the time of the release; people with chemical engineering backgrounds had been replaced, apparently to cut operating costs at the money-losing plant.

Only one corporate HQ safety audit is known to have been done over seven years of Union Carbide India plant operations; no check had been done in the two-and-a-half years since 1982.

The community did not know how dangerous the plant materials were; some neighbours thought it was a medicine factory. This stands in contrast to the USA, where West

Virginia has recently enacted a right-to-know law over industry objects; in New Jersey, where the strongest state right-to-know law is being challenged by industry, Union Carbide is leading the attack in court and in the state's advisory council.

India has no meaningful government regulation regarding air pollution and occupational safety and health; even the 'requirement' that air pollution engineering controls be operational was ignored by the plant management and not cited in occasional, perfunctory inspections by the government.

There is no government regulation of zoning at all in India, although Union Carbide had been asked to move the plant outside town even before it was opened (1975) by an official who was soon transferred to another position.

Cumbersome Indian laws for compensation for wrongful death and personal injury have even been cited as the basis for US courts allowing a lawsuit to be tried in the USA over an accident that occurred in India. Hardships on plaintiffs include:
☐ high filing fees (£300 to sue for £10,000), no possibility of getting representation on a contingent fee basis;
☐ slow process of getting cases to trial.

Other questions include:

Why did Union Carbide use the more dangerous MI route to make carbaryl pesticide in a country like India, without even doing frequent safety audits after atrocious conditions were found in 1982?

Why did Union Carbide India not follow the example of Bayer in Germany, which manufactures only small amounts of MI at a time for immediate use (hence no large-scale storage of MI)?

BHOPAL: BACKWARD OR ADVANCED?

TARA JONES

Attempts to dismiss the Bhopal massacre as an incident that could occur only in a Third World country were as certain a response from international toxic capital as was the Union Carbide Corporation's attempt to transfer all blame for the massacre to its Indian subsidiary. Thus Martin Trowbridge of the UK's Chemical Industries Association, explaining how Bhopal could not occur in England, spoke of the 'culture of safety' which is lacking in the Third World. An anonymous international economist confided to the *Los Angeles Times* that 'the level of technical expertise among workers in developing countries isn't anywhere as good as in developed countries.' Lee Talbot, visiting fellow at the World Resources Institute, said that disasters such as Bhopal could become increasingly common in industrializing countries that lack the trained workers and government regulators to find and correct hazards in larger and more complex plants. The *International Herald Tribune*'s first two conclusions on Bhopal were that 'Hazardous facilities often pose added risks in developing nations, where skilled labour and public understanding are often lacking. Special training is needed', and 'public education is critical in developing countries, where people often do not understand the hazards of toxic substances.'

In response to this depiction of Bhopal as a result of bad, unskilled workers in an underdeveloped country let loose on a highly complicated and dangerous technology, critics have pointed to whose responsibility it is to train the workers and design the plant safely. This response is good as far as it goes, but it does not go far enough. True, UCC did dump obsolete technology on Bhopal, cut back on skilled staff there and abandon the original safety measures. At the same time, however, both the background and the dynamic of the Bhopal disaster bear similarities to those of more technologically sophisticated disasters.

Something more was involved in Bhopal than bad technology combined with unskilled workers in an environment unsuited to industrial culture, as was shown by the leak at UCC's Institute (West Virginia) plant the following August. Here we had 'new, improved' chemical technology, a highly skilled workforce and an environment dominated by industrial culture. Yet the Institute leak seems to mirror many aspects of the Bhopal massacre, with the exception of the horrific death toll.

As we will see, the problem most horrifyingly revealed at Bhopal lies deeper than the partial response mentioned above would suggest; that is, the problem is rooted in the nature of the chemical technology itself. I use 'chemical technology' here in a wider sense than simply the pipes and process lines, the vats and scrubbers, the flare stacks and storage tanks. For,

> just as capital has been confused with the industrial apparatus and accumulated wealth, when in reality it is more than industrial plants and wealth, but also social relations, so has technology been confused with machines and tools, when it is in reality a complex of social relations, a 'web of instrumentality', a qualitatively different form of domination. (T. Fulano, *The Fifth Estate*, July 1981)

Thus the social relations of command and control are as essential to an understanding of what went wrong as are the mechanical details of failed systems.

I would like to support this argument by examining briefly the accident on Sunday 11 August 1985 at UCC's Institute plant, showing the similarities between that accident and the Bhopal massacre and briefly noting similar aspects of the technology.

Bhopal: Backward or Advanced? 285

The leak at Institute released approximately 2,800lb of aldicarb oxime (AO) decomposition products, approximately 700lb of methylene chloride and some 500lb of residues. The leak occurred in a new chemical process begun in May 1985 so that UCC would not have to ship large amounts of MIC to its plant in Georgia for formulation into the pesticide Temik. Thousands of local residents were trapped inside their homes for two hours. One hundred and thirty people were treated at hospital for eye, nose, throat and lung burns; thirty were hospitalized overnight. One of the six UCC workers injured was reported to be in a serious condition with eye injuries.

The most immediately obvious similarities between Bhopal and Institute are the similarities in the accident itself and in the fact that the safety system was unable to contain the chemicals. Thus, as in Bhopal, there appears to have been a runaway reaction, leading to a rise in pressure which burst open three gaskets on a tank containing AO. The escaping chemicals were vented to a scrubbing tower (for neutralizing chemicals) and a flame tower (where chemicals not neutralized would be burned). This safety system was overwhelmed by the amount and pressure of chemicals flowing into it. While some chemicals were neutralized, a rupture disc on the safety system was blown by the pressure, so more chemicals escaped through the safety system. Thus pressure on the safety system was higher than the pressure the system was designed to handle. To quote the *New Scientist* account of the leak, 'the system coped with a runaway reaction little better than equipment at Bhopal, even though it had been installed as a direct result of that accident.' In Bhopal also the safety system was overwhelmed by the pressure of escaping chemicals. The Institute works manager said that, as at Bhopal, there was a water spray system near the tank to contain leaks, but it was not strong enough to deal with the AO leak.

Similarities with Bhopal do not stop there. Local community officials charged that UCC took hours to reveal what chemicals had leaked, thus making early medical treatment into a guessing game, as at Bhopal. Even when information was provided, UCC downplayed the hazards. Dr Jack Tolliver, chief of emergency medicine at Charleston Area Medical Center said that UCC's doctor at the Institute plant told him on the day of the leak that AO was 'a very minor irritant' that would have 'no long-term effects'. Similarly, UCC said that AO was not as toxic as MIC, the gas that had leaked at Bhopal. Yet Jerry Dodson, a lawyer for the US House Energy and Commerce subcommittee, released a memo (dated 28 November 1983) which UCC had submitted to the subcommittee: in it UCC had categorized AO, like MIC, in the most toxic of the four categories by which UCC classifies the chemicals it uses. According to that UCC memo, chemicals in that category can cause cancer, birth defects, genetic damage, irreversible nervous-system disorders and immediate serious illness. The UCC doctor at Institute with whom Dr Tolliver spoke on the day of the leak had been sent a copy of that memo. According to Dodson, UCC's contention that AO is less toxic than MIC 'completely misstates what Union Carbide told us in 1983.'

Again, there was a startling lack of information on AO. Federal officials said that they had little or no information on AO's effects on people's health, as it is a rare chemical. An EPA official said, 'When it comes down to an intermediate, we don't regulate it. Union Carbide probably has the best information.' UCC said that it did not believe the injured would suffer residual effects, but that the company had no plans to monitor the people exposed to find out.

UCC spokesperson Thad Epps said that among the new safety systems installed at Institute was a computerized emergency response system that automatically informs local emergency services of a leak and monitors weather conditions to predict which direction the gas would take. 'The system worked perfectly,' Mr Epps said. According to a company statement, 'The emergency response system worked according to plan, and all government authorities and hospital emergency rooms were notified in a timely fashion.' Plant officials said that the rescue services and hospitals were warned 'about five minutes' after the gas was detected.

Alas, it was not so. The mayor of nearby Charleston said, 'The system didn't work. Our communications centre wasn't notified.' The mayors of South Charleston, St Albans, Nitro and Dunbar said that they learned of the leaks only through news broadcasts. UCC said that it delayed informing Kanawha County officials of the leak for twenty minutes as the toxic cloud appeared to be hovering over the plant. Also, its computerized model said that the toxic plume would not threaten the local community.

From Sunday to Monday UCC's timetable of the leak changed. On Sunday UCC said that the leak began at 9.35, the siren was sounded at 9.40 and the leak ended at 9.50. On Monday the company said that the leak began at 9.24, emergency officials were notified by 9.44 and the siren was sounded by 10.00. It is worth noting that local residents say they saw the plume rising from the plant at 9.15.

There is a grotesque irony in the fact that the new safety features attached to the plant made their own contribution to the accident. Incorrect results from the much-publicized computerized equipment persuaded management that the leak would not spread beyond the plant site. These incorrect results came from the fact that the computerized alarm system suffered from a 'data deficit' on AO. Thus the model of the toxic plume's behaviour was based on information relating to MIC. Indeed, it was later established that the system was programmed for only three toxic chemicals – phosgene, chlorine and MIC. A UCC spokesperson said that AO was not programmed for, as the company placed it lower down the scale of toxicity and volatility than MIC, phosgene and chlorine.

As at Bhopal, people learned first of the escaping toxic gas through the evidence of their own senses rather than through any new, improved, computerized warning system. The developed infrastructure, skilled workforce, industrial culture and technologically advanced modelling and warning systems intended to 'make a safe plant safer' led to a situation where one woman who noticed the pungent smell 'thought it was the cat litter. But then I opened the door to pick up the paper and it almost knocked me down.' The situation among well-trained workers was as follows:

288 Anti-Racist Science Teaching

> The gas clouded up the room so badly that the workers couldn't see. They could only find one respirator, so they laid on the floor and used it among the six of them until a rescue unit came. There was no warning, no horn, no pressure indicator, nothing.

That was Gerry Robinson of the International Association of Machinists talking. He also said:

> But there are many questions to be answered. There should have been an indication of what was happening on the pressure gauges; there should have been a warning horn sound; and there should have been a trip valve to empty the gas into a scrubber. There were supposed to be all these safety measures, monitored by computer, and still something failed.

A lot of things failed. According to UCC's own preliminary investigation, there had been several occasions where standard operating practice had not been observed in the eleven-day-long sequence of events that led up to the gas leak. Among these, using a reactor vessel as a storage tank for AO is not standard procedure 'but not unheard of'. UCC's study of the spill said that from the first of August until the leak occurred, steam flowed from a set of leaky valves into the storage tank which contained 4,000lb of AO and methylene chloride. None of the twenty workers and five forepeople who work near the AO tank noticed any problem. A standard safety check had not been made on the tank, which was not designed to hold AO and had a malfunctioning pump which contributed to the leak, while the tank itself had not been in use since November 1984. OSHA said that the accident came from poor design and human error, while EPA said that the plant's safety procedures were inadequate.

Another basic similarity between both accidents is that they may be traced back to decisions based on capital's values of profit-taking and cost-cutting. If the major cause of the Bhopal massacre can be traced back to the original export of unsuitable and ill-designed plant and the 'rationalization' of staffing levels in response to the plant's lack of profitability, then a similar economic motive can be found for the Institute leak.

The MIC plant at Institute was immediately closed after the Bhopal massacre. This was necessary to back up UCC's claim that essentially the same level of safety existed in the US as in India. At the time, UCC said it would not reopen the plant until it knew for certain what had gone wrong at Bhopal. UCC's March 1985 report on Bhopal left many questions unanswered – including the source of the water contamination that precipitated the accident – and was condemned by US environmentalists as highly speculative. Nevertheless, despite these unanswered questions, UCC said it saw 'no reason' why the Institute MIC plant should not be reopened.

And there were plenty of reasons for UCC to reopen it. UCC indicated that it might incur further charges against earnings if the Institute plant was not reopened by 1 April as planned. By the second week in April, when pesticides were needed for planting starting that week, a major shortage of MIC-based pesticides was reported due to the Institute plant shutdown. (These MIC-based pesticides make up nearly a quarter of all pesticides produced in the world.) The President of Union Carbide Agricultural Products Company estimated the shortage of these pesticides would average 10–30 per cent. In response to the shortage, prices of UCC's MIC-based pesticide Temik had risen 20 per cent. These shortages could not be made up by increasing production later in the year, so reopening of the Institute plant was essential to prevent users from switching to alternative products. Temik is a key component of UCC's agricultural operations, which account for 5 per cent of UCC's sales.

So, on 4 May, UCC reopened the plant after detailed government inspections and repeated assurances from UCC that the plant was safe. UCC had spent $5 million on installing new safety devices, including the much-publicized computerized emergency response system. Before reopening of the plant, UCC invited 800 residents who live nearby to attend meetings. The press was barred from these meetings and none of West Virginia State Colleges's 4,000 students, some of whose dormitories are a few hundred yards from the plant, was invited. In declaring the plant safe, West Virginia state and Federal regulators showed the same optimism towards Institute as did the Indian state authorities towards Bhopal. OSHA, EPA and local officials also blessed the plant as safe.

The difference from Bhopal is that the same optimism was shown by workers and their representatives, business or otherwise. James Miller, a distribution specialist who handled MIC at Institute for sixteen years, was quoted by the *New York Times*: 'The news media and the politicians have blown this way out of line. There's nothing dangerous here.' Mr Gresham, 'business representative of Local Lodge 856 of the International Association of Machinists and Aerospace Workers', said he knew of 'no leaks that could not be handled by the workers on the job'; however, regarding the EPA report of twenty-eight leaks in Institute from 1980 to 1984 released that morning, he admitted, 'I don't know if the leaks were major, minor or whatever. This morning was the first I ever heard of it.' Even the West Virginia Citizens Action described previous MIC releases (not leaks, 'releases') at Institute as 'routine'.

Providing another parallel with Bhopal, the IAM have claimed that government inspection of the plant was inadequate. According to George Robinson, IAM health and safety director,

> We asked the government to be more rigorous in their inspections and to monitor the plant on a continuing basis. We felt that the unit handling MIC was treated differently from other parts of the plant where safety checks were not carried out as they should have been by inspectors from the Department of Labour.

Finally, those who felt that at least after the Institute leak UCC would get its act together cannot have been reassured by the spillage of hydrochloric acid at UCC's South Charleston plant on 27 August. According to Lieutenant Larry Mullins of Kanawha County Sheriff's Office, 'We had a problem with the line of communications between Union Carbide and our emergency communications centre. It took some time before

Bhopal: Backward or Advanced? 291

we actually received word of it.' One and a half miles upstream, 60,000 people were listening to Chubby Checker: none of them appeared to have heard the alarm whistles or to be aware of the chemical spill.

Coverage of Bhopal has spoken of ignorant Third World workers, as though their ignorance was exceptional. Yet this ignorance is the norm. The reluctance to give relevant information to workers or the community is a central characteristic of toxic production in both metropolitan and peripheral countries. The most obvious explanation for this is that details of chemical processes are trade secrets, jealously guarded by companies as wealth-producing information. After all, information is an instrument for power and profit; in such circumstances, for the West the 'free flow of information' is just one more myth. This ignorance is worsened and facilitated by the increased use of computerized process-control systems, which leave plant workers as machine-minders and gauge-readers. Finally, of course, it is helpful for toxic production if workers are kept unaware of the toxic nature of the work process imposed on them.

To see the prevalence of this limitation on information on the occupational hazards of toxic chemicals 'even' in such a safety-conscious, industrially developed country as England, you need only examine Maurice Frankel's *A Word of Warning* (Social Audit, 1981). Perhaps more useful in illustrating this advanced industrial culture of ours is the graphic example of workers at Rechem International's toxic waste incinerator. 'Spillages are common', said one, while another asserted, 'minor blasts and explosions are regular occurrences' and, as in Bhopal, 'workers are branded as troublemakers if they complain about safety standards.' Lack of information from the company can be exemplified by one worker whose lack of safety in handling PCBs can be as easily attributed to ignorance as can the Bhopal workers':

> We used to split open the metal containers with picks. The PCBs went everywhere. It used to splash over our overalls and bodies. We thought nothing of it. We treated it like oil.

The company never warned us. We only found out what we were dealing with when a shop steward looked up the meaning of PCB in a medical dictionary.

Similarly, workers at Merck, Sharp and Domhe's plant in Tipperary (Ireland), which is the centre of a major pollution controversy, admitted at meetings held by local farmers that they knew next to nothing of the materials they handled at work. Despite this, the Merck management were able to produce their hazards information sheets at the later High Court trial and inform the judge that they were readily available to workers.

The ignorance is not confined to workers. In an editorial after the Bhopal massacre, the *New Scientist* pointed out that the whereabouts of the 1,500 factories designated as hazardous in the UK are secret, as is information on their risks and whatever plans exist to deal with emergencies at these plants. And the Chemical Industries Association, which is so proud of the culture of safety in the UK, wants to keep it secret. While the Health and Safety Executive had intended to name those 1,500 plants in accordance with the EEC Seveso directive on major hazards, the Confederation of British Industry and Mr Trowbridge's Chemical Industries Association opposed this disclosure, with the result that only 250 sites will be covered by the planned HSE regulations.

As we have now seen, a similar ignorance of hazards exists – or, more accurately, is created – in both metropolitan and peripheral countries. Therefore, when Western newspapers editorialize on the need to educate ignorant Indian workers and people in the hazards of toxic processes, they would do well to prescribe such education in their own countries. The difficulty of getting information on toxic industry in the 'developed' as well as in the 'underdeveloped' countries may become a major area of struggle. Racist explanations of the massacre at Bhopal are a veil behind which toxic capital attempts in vain to hide its responsibility.

(Written autumn 1985)

Official Documents and Commentaries

INTRODUCTION

This book has emphasized the actual practice of science and of science teaching, rather than declarations of good intentions. At the same time, of course, official policy does affect teachers' willingness – or not – to risk transforming that practice. In this section we present some key policy documents and critical commentaries.

In 1983 the Inner London Education Authority (ILEA) became one of the first in Britain to adopt an anti-racist policy, part of which we reprint here. Significantly, the statement criticizes both subtle and overt forms of racism, as well as educational selection according to stereotyped views of ability. However, by failing to suggest any connection between racism and class divisions, or racism and imperialism, it lends itself to interpretations that attribute racism simply to personal prejudice.

ILEA's anti-racist policy has been virtually ignored by science teachers, some of whom have even argued that it did not apply to science. Alarmingly, though perhaps not surprisingly, a similar argument came as well from ILEA's Staff Inspector for Science himself. Basically, he argued that Western science is already multicultural anyway, by virtue of holding universal validity. In one of many critical replies, a head teacher pointed out the crucial scientific contributions of non-Western civilizations. Eventually the Inspector attempted to cover himself by issuing a clarification; in effect this sustained his original position by making false distinctions between science and its history, and between 'pure science' and its applications.

ILEA's Chief Inspector, David Hargreaves, previously convened the commission that produced *Improving Secondary Schools*, known as the Hargreaves Report. Claiming to deal with the problem of working-class 'underachievement', the report virtually ignores racism. And while acknowledging the role of public examinations in demotivating students, its proposed remedies would seem to compound the problem. Here we include a brief critique.

Lastly we move to the level of central government. Its landmark document, *Science 5–16*, expresses the laudable intention of making science more accessible to students. Yet its national guidelines simply reinforce the traditional aims of science education, while possibly narrowing the space available for teachers to develop anti-racist approaches. Here we present a critique written out of discussions of the Association for Curriculum Development in Science.

ILEA ANTI-RACIST STATEMENT

1. The Inner London Education Authority is committed to achieving equality in education and employment in the Inner London education service. This means the development of an education service from which racism, sexism and class discrimination and prejudice have been eliminated so that the Authority can respond fully to the needs of our multi-ethnic society. This paper is concerned with one of the three major obstacles to achieving equality – racism.

Many reports, official and unofficial, have indicated clearly the extent and effects of racism in education. The chief victims are black people, i.e. Afro-Caribbean and Asian communities. Other ethnic minorities are also subjected to racial prejudice and discrimination. However, it is in the interests of all our employees, students and pupils that we actively seek to eliminate racism in all our institutions and in all branches of the service.

There is, rightly, among the black communities and other ethnic minorities an implacable opposition and resistance to racism. This is a powerful and positive factor in British society. Another force is also available to the service. This is the strong tradition in British society of opposition to injustice in whatever shape or form. All employees of ILEA and users of the Authority's service are uniquely placed, if only they would seize their opportunity, to educate generations of young people free of racism and prejudice.

2. It is necessary that all those who serve the Authority, or

benefit in any way from its services, understand what is meant by the concept of racism.

In structural terms, racism is represented mainly by the formula: racism = power + prejudice and discrimination. Accordingly, power and resources in the education service, as in other institutions of our society, are in the hands of white people. It is essential to understand that dimension of the power relations between black and white people because of the part it plays in sustaining inequality at the national as well as at the international level. Other explanations of the concept are necessary if we are to get a total picture of what the phenomenon represents.

Racist ideology, surrounding attitudes, values and beliefs, is based on the assumption that black people are inferior to white people. Embedded deeply in the procedures and practices of our institutions are many such notions. Some individuals and groups within the service may act in a prejudiced and discriminatory way, deliberately or because of such unexamined beliefs.

Given that we live and work in a society where racist practices and attitudes permeate the whole system, there are several ways in which we can be racist:

(a) We can be openly hostile to ethnic minorities on grounds of race or colour.

(b) We can claim to see no difference between white people and black people or other ethnic minorities, and thus deny the significance of racism as a factor in sustaining inequality between black and white people in our education service.

(c) Even those of us who have made every effort to rid ourselves of racism may fail to see how deep-seated racist attitudes which have been prevalent in our society for so long affect our treatment of ethnic minorities.

(d) We may fail to see how decisions we make, however fair and valid according to the traditions of our system, affect ethnic minorities adversely because of the basic inequality in our society.

The most significant manifestation of racism is the absence of black people and other ethnic minorities from positions of power and decision-making in our service. Every institution and branch of the service must take active steps to redress this imbalance and to ensure that the perspective of black people influences policy-making directly.

Racial discrimination represents the other most obvious manifestation of racism. One form of this is overt, i.e. deliberate with intent, and blatant. It includes such practices as personal abuse, graffiti, provocative behaviour and, more seriously, racist attacks by one group on members of another.

Other forms of discrimination are less easy to perceive but are by no means less important. They include procedures employed within the education service, in its administration as well as in its institutions, which, however well intentioned or rooted in custom, have the effect of reducing the opportunities open to members of ethnic minority groups. These procedures need to be re-examined and altered to take into account the Authority's priorities. Similarly, there are deep-seated attitudes which affect adversely members of ethnic minority groups. These attitudes need to be identified, brought to the surface and openly challenged.

3. It follows that every aspect of the Authority's work, every branch and every institution, must be examined with the clear objective of eradicating racist practices and assumptions. The responsibility for developing and implementing the Authority's anti-racist policies must rest with every employee in every institution and every branch of the Authority's administration. It is for every institution and branch of the Authority, therefore, to develop a coherent action programme for the elimination of racist practices.

The ILEA has a straightforward legal obligation to eliminate racial discrimination. That is to say, the Authority and everyone who works within it are obliged to comply with the provisions of Section 71 of the Race Relations Act 1976. That section places a duty on every local Authority to:
(a) eliminate racial discrimination;
(b) promote equality of opportunity, and good relations, between persons of different racial groups.

The educational duty is equally important. Following its initial Multi-Ethnic Policy Statement in November 1977, the Authority has consulted widely with community groups, teachers and heads and is currently taking new initiatives designed to eliminate racism from the education service.

4. In giving expression to its commitment to combat racism, the Authority asks:
(a) that all educational establishments, through their staff and governing or managing bodies and in association with the committees they serve, prepare and publicize carefully thought-out statements of their position. This must be seen as part of the Authority's legal and educational commitment.
(b) that annual reports for governing bodies, as part of the regular process of keeping schools, colleges and other establishments under review, should incorporate information on what action has been taken on curriculum, staffing and organization to meet the needs of a multi-ethnic society.
(c) that all administrative and other branches of the Education Service should re-examine their procedures to ensure that, however unintentionally, they do not have the effect of discriminating against members of ethnic minority groups.

The Authority itself is committed to eliminating racism and to take such action as it properly can to remedy its effects. Its success in doing so will depend upon the determination of every individual within the education service to ensure that appropriate action is taken. In issuing this statement, the Authority reaffirms its own commitment to promoting equality of opportunity, and good relations, between persons of all racial groups.

ELEMENTS IN THE WORK OF THE SCHOOL OR COLLEGE (Excerpts)

There is a range of elements in the life and work of the school or college which can be positively employed in order to tackle racism and so improve education.

CONTENT AND ORGANIZATION OF THE CURRICULUM AND RESOURCES

Teachers have a great deal of control over the content of what pupils and students learn and the resources that are made available. Through their selection of content and resources they can therefore make a positive contribution.

(a) Very young children need both an affirmation of the value of people of all colours and cultures and to be helped towards avoidance of stereotypes and misrepresentations which form at a very early age.

(b) A wide range of content is important, but it is essential that pupils develop analytical skills and can engage in an understanding of cross-cultural perspectives and values.

(c) Pupils and students must have opportunities to gain an historical perspective that is free from ethnocentric biases.

(d) The whole curriculum must be open to all so that no sort of restricted access is given to some pupils because of stereotyped views of ability.

PROCESSES OF SELECTION AND GROUPING

In the past ten years, the education system has been criticized for discriminating against certain pupils and students by unfair processes of selection and grouping. This can occur where there are 'nurture groups', banding or streaming, or withdrawal groups, and in the selection of examination groups in secondary schools:

(a) It is essential to ensure that no selection of these kinds is affected by conscious or unconscious stereotyping of an ethnic group or black pupils.

(b) It is also important to match any special provision with a clear and accurate diagnosis of why special provision is thought to be needed.

(c) As has been stressed earlier, knowledge of pupils and students must be free of any false notions of inherent ability based on ethnic or cultural diversity.

ILEA CORRESPONDENCE

A LETTER TO HEADS OF SCIENCE IN ALL ILEA SECONDARY SCHOOLS

I gather that you may soon be asked to produce a statement on the attitude of your department to multi-ethnic issues.

In doing this, you may be helped by the following extract from a statement which I recently produced on behalf of the science inspectorate:

Although the pure and applied aspects of science are inextricably linked in any satisfactory presentation of the subject, it is helpful in this brief statement to deal with them separately.

1. PURE SCIENCE

It is a truism which needs no emphasis that pure science is entirely multicultural and that the community of scientists transcends national and ethnic boundaries. Consequently, the teaching of science *per se* is ethnically and culturally neutral: scientific facts, principles and laws apply in a uniform manner in all societies and in all parts of the world. As a result, there are few if any scientific concepts or terms to which anyone could reasonably take exception on 'ethnic grounds'. Science inspectors and science teachers are, however, very much aware of the need for a sensitive handling of a few topics – particularly biological ones, such as blushing and the incidence of sickle cell anaemia.

Most illustrations in science textbooks and readers are of apparatus, graphs, etc. In many published books, however, the children and adults who may be pictured are usually white Caucasians. It must therefore be emphasized that the organizers of the various ILEA science curriculum projects (Insight, APPIL, ILPAC and ABAL) have taken great care to show a variety of ethnic types in their printed materials.

Virtually all the science taught in schools originated in Europe or North America. Consequently, it would be dishonest (not to say impracticable) to suggest to children that modern science derives equally from people of all backgrounds. In this connection, however, mention might be made of the wholly disproportionate contribution of Jewish scientists.

2. APPLIED SCIENCE

Good science teachers take every opportunity of quoting relevant applications, no matter what kind of science they are teaching and to whom they are teaching it. Nevertheless, applied rather than pure science is of particular importance in science courses of a general rather than a specialist nature. And in this context, ILEA science teachers are urged to lose no opportunity of discussing such matters as the provision of a water supply in different parts of the world, building materials appropriate for different climates, natural resources, alternative and intermediate technologies, etc.

Dr John Spice
Staff Inspector of Science, ILEA
15 September 1983

REPLY FROM AN ILEA SCHOOL

SCIENCE AND MULTI-ETHNIC ISSUES: SOME REMARKS ON DR SPICE'S NOTE

Dr Spice says that 'pure science is entirely multicultural ... Consequently, the teaching of science *per se* is ethnically and culturally neutral: scientific facts, principles and law apply in a uniform manner in all societies and in all parts of the world.' This seems to ignore two aspects: first, even in so-called 'pure' science, the postulation of a hypothesis by a scientist, who is a human being with a certain cultural and ethnic background, cannot be totally culture-free; second, it is now accepted that scientific 'observations' cannot be divorced entirely from the subjectivity, the percepts, of the observer, and another cultural element therefore enters here.

Dr Spice, still under the heading of 'Pure Science', further says that 'virtually all the science taught in schools originated in Europe or North America'. The important word here is perhaps 'originated'. One could and should draw attention to origins of astronomical study in the river civilizations of Mesopotamia and Egypt, closely connected with the development of mathematics. For 500 years after the collapse of the Roman Empire, too, the centre of scientific advance was east of the Euphrates – in Persia, Syria and India. (Perhaps Dr Spice would separate science from mathematics, but one ought to point to the development of the modern system of digits and especially of the zero, which made proper use of place-value possible, in India, and to its transmission to Europe by the Arabs.)

Dr Spice says that pure and applied science are so inextricably linked that separation is unsatisfactory. One might point then to the stainless iron pillar of Delhi (c. 400 AD), to the development of the astrolabe by Abermi (an Arab) in 1060, and the use of crystal lenses by Alhazen in 1038, with the study of refraction, reflection and the use of spherical mirrors around the same time, leading to a theory of light rays. (Bacon based much of his scientific method work on all this.) Under the Caliph al Mamun in 830 AD a degree of latitude was measured twice on the shores of the Red Sea; for 800 years after that, European official dogma stated that the earth was flat. Al Jabir began methodical study of chemistry, with distillation, crystallization, and filtration. Soon general principles of classification of substances followed (alums, alkalis). Do we need to point to words such as 'algebra', 'nadir', 'zenith', 'azimuth' as evidence of Arab pre-eminence? The Arabs did much more than merely transmit Greek science to Europe. In medicine, botany, the study of herbs and in trigonometry the Arabs were original thinkers and discoverers. This casts fresh light on Dr Spice's assessment of 'the wholly disproportionate contribution of Jewish scientists'.

Nor would one ignore the Chinese contribution. Dr Needham's vast work on Chinese science in the past two millennia is sufficient indication; everyone knows that paper and gunpowder were Chinese inventions.

It is also worth pointing out that certain sciences (notably, genetics and psychology – the latter especially in its application to psychometry) can be used in a racist way.

Dr F. D. Rushworth
Head Teacher, Holland Park School
19 October 1983

SCIENCE AND MULTI-ETHNIC EDUCATION

PUBLISHED AS A SPECIAL NEWS BULLETIN, SENT TO ALL ILEA SECONDARY SCHOOLS

1. On 15 September 1983, I sent you a note of some of the matters you might wish to take into account when you came to prepare the contribution of your department to your school's statement on multi-ethnic education. That paper was perhaps over-concise, and in some respects it was incomplete. If, therefore, you are intending to use it, I hope you will do so in conjunction with this present paper. Last term's paper and this, taken together, are endorsed both by my colleagues in the science inspectorate, and by the members of the multi-ethnic inspectorate, as a basis for further discussion of the various issues which are involved.

2. In particular, I now see that the two sentences: 'Virtually all the science taught in schools originated in Europe or North America. Consequently, it would be dishonest (not to say impracticable) to suggest to children that modern science derives equally from people of all backgrounds' do not clearly convey what I mean to convey. I regret any confusion this may

have caused, and I ask you to replace these two sentences by the following statement:

Peoples from all parts of the world and at all periods have contributed to scientific knowledge. But during the period when modern science became established (roughly from 1550 to 1900) conditions for scientific work were, for a number of reasons, far more favourable in Europe (and, in the nineteenth century, in North America) than elsewhere. With the progressive establishment of education for all in every country, predominance of Europeans and North Americans in scientific work has become increasingly less marked. And this trend will certainly continue. Every effort should be made to leave children in no doubt about these facts. However, the difficulties of doing so in a way which will carry conviction must not be underestimated. On the one hand, the wholesale revision of school science syllabuses in the 1960s effectively eliminated most of the historical background. For instance, the typical introductory chemistry book of twenty-five years ago probably included a picture of an Arab alchemist, whereas nowadays it may well contain little or no reference to individual scientists. (But the origin of the term 'alkali' should certainly be explained.) On the other hand, most of the advances in science of the past century have, obviously enough, concerned matters which are not taught at school. Thus, the fact that the existence of mesons was postulated by the Japanese theoretical physicist Yukawa in 1934 would not normally be mentioned, since particle physics does not form part of school science syllabuses. Because, for the most part, *school* science is a product of Europe and North America, it is all the more important to emphasize, as occasions arise, that modern science might well have developed in (say) Africa rather than in Europe and North America had conditions been different, and that science, as an organized body of knowledge, is the inheritance of all peoples. If this idea (although it may to some be self-evident) can be conveyed to all pupils in a natural (and hence convincing) manner, this in itself will represent a most significant contribution of school science to multicultural education.

3. It is agreed that pure science has universal validity and as such is entirely multicultural. On the other hand, access to knowledge and control of the applications of science are determined by social, political and economic factors. The consideration of such matters is largely absent from examination syllabuses, if only because of the sheer mass of pure and applied science which has to be taught. However, the discussion of these aspects offers far more opportunities for demonstrating the multicultural nature of science than does (say) an enquiry into the properties of aldehyde and ketones, or an examination of the characteristics of an electromagnetic field, or a study of the amoeba. Despite the fact that the discussion of these factors may take place more obviously in courses like 'Science in Society', no opportunity should be lost to discuss them at appropriate points in *all* science courses.

4. In addition, occasions do arise in the teaching of school science (although less frequently than in many other subjects) for pointing out how particular topics may be perceived in different ways by men and women of different cultural backgrounds. It is particularly hoped that science teachers will take full advantage of all such opportunities of furthering the aims of multicultural education.

5. It is of paramount importance to combat the kind of prejudice which is based on pseudo-scientific arguments which are in fact spurious – leading to myths such as a supposed correlation between innate intelligence (if there is such a quality) and race.

6. A rather different point is that special efforts should be made to ensure that pupils from all ethnic groups are motivated to continue to follow appropriate science courses in and above the fourth secondary year.

Dr John Spice
Staff Inspector of Science, ILEA

THE HARGREAVES REPORT: MONOCULTURAL EDUCATION?

The Hargreaves Report, entitled 'Improving Secondary Schools', was published by the Inner London Education Authority in 1984. The report presents the findings of a committee chaired by Sir David Hargreaves, who subsequently became ILEA's Chief Inspector of Schools. Having attempted to define a broad consensus on the problems of underachievement and indiscipline, the report was greeted as a progressive response. It can be expected to affect educational policy and practices in London, though its influence is likely to extend far beyond one city.

For those reasons it is important to analyse its premises, findings and proposals. This brief critique indicates the report's failure to identify the source of the problems that it proposes to solve. Far from challenging institutional racism, the Hargreaves Report may well help to perpetuate it.

'UNDERACHIEVEMENT'

The report focuses on 'secondary schools as they affect pupils mainly in the age range eleven to sixteen, with special reference to pupils who are underachieving and those who show their dissatisfaction with school by absenteeism or other uncooperative behaviour'.

As remedies, the Hargreaves report favours 'group work, co-operative learning and encouraging pupils to find out for themselves', 'the development of lively and enquiring minds' and 'the ability to question and argue rationally'.

Among the proposed aims for the secondary school curriculum are:

☐ to acquire understanding of the social, economic and political order and a reasoned set of attitudes, values and beliefs; and
☐ to help pupils to understand the world in which they live and the interdependence of individuals, groups and nations.

It would therefore seem that the Hargreaves Report advocates some thought and discussion of social, economic and political matters. Unfortunately, the compulsory curriculum recommended in the report does not include any humanities subject. History, geography and economics are listed among the optional subjects together with languages, physical education and commercial and business studies. Students may choose to take as their optional subjects additional science, English, mathematics or 'aesthetic' and 'technical' subjects which form part of the compulsory core. In the core, three periods a week are allocated to personal, social and religious education.

With such a curriculum structure, it would be difficult to implement the report's proposed aims. Many of the important issues would most likely be relegated to 'Personal and Social Education', which would comprise only 7 per cent of the timetable. Many of the issues would be neglected or be submerged amongst other topics such as careers, pastoral work, study skills, religion and health – that is, unless the approach to those topics is radically changed to include social, economic and political aspects.

The report says surprisingly little of relevance to ILEA policy statements which propose to rid the school system of racism, sexism and inequalities based on social class. Regarding black 'underachievement', it simply recommends that ILEA 'continue with research to uncover the factors associated with high achievement among Afro-Caribbean pupils', in order to strengthen those factors. In virtually ignoring an issue of such major importance to all students, the report could be seen as actually racist by omission – or even by commission. Indeed, it uses a model of cultural deprivation in order to attribute underachievement to individual students' insufficient motivation, in turn supposedly due to cultural deficiencies of their families.

The report does acknowledge several sources of student disaffection, such as the pressure of exams, inappropriate or irrelevant curriculum, and a feeling of humiliation from not understanding a lesson. When pupils experience their classroom situation as a threat to their sense of motivation and self-worth, 'it becomes rational for them to play truant or to protect themselves by classroom misbehaviour.'

The report also recognizes that 'the present system of public examinations guarantees underachievement and disaffection' by excluding many social skills from assessment and by in effect excluding some pupils from all public examinations.

However, the report's proposed remedies – such as individual profiling and graded assessments – can hardly overcome those sources of disaffection. Least of all can they reduce the power of competitive public exams to pressurize and label students. Some teacher retraining and smaller class sizes would certainly facilitate the curriculum changes needed to encourage rather than demotivate students. Yet the report simply accepts that 'resources will remain stable or even decline'.

Noting that Afro-Caribbean parents have set up supplementary schools for their children, the report expresses concern 'that their existence may reflect a dissatisfaction with the ILEA secondary school system'. Yet it offers no analysis of why those parents felt dissatisfied, nor why their children should feel more motivated to learn in those schools than in ILEA's. By failing even to search for the reasons, the report in effect absolves ILEA of responsibility for providing adequate education for *all* children; this makes it complicit in institutional racism.

RACISM

Furthermore, the report ignores the influence of racism on the social, emotional and intellectual development of children, black and white. It says nothing about the racism which pervades the curriculum, much less the need for anti-racist education. Indeed, it reduces racism to a problem of 'racial prejudice' by individuals who bully, taunt or ridicule 'racial minorities'; thus it implicitly absolves the schools of institutional racism.

Ironically, the report suggests that merely teaching about racism as curriculum content may be controversial:

> Parents expect to be informed on curricular matters where they have very strong views or the content is potentially controversial – for example, health and sex education (where their right is, in fact, a legal one), racism and sexism.

Yet it nowhere suggests that teachers consult parents or students about the effects of racism in the running of the school; much less does it offer support to anti-racist teachers.

More generally, it fails to take seriously racism, sexism and class divisions as major causes for the 'disaffection' that it proposes to overcome. Instead, it takes for granted the school's official culture of academic success and proposes ways to integrate 'underachieving' students into it. For example, almost as a technique of behaviour modification, more frequent and accessible achievement tests are supposed to enhance such students' motivation, apparently defined as willingness to compete for success on standardized tests. These students are expected to overcome their cultural deficiencies by identifying with the white middle-class values of the school.

Accordingly, when the report says that 'pupils work better when parents and school work better', it does not mean that the school must validate and incorporate the values of working class and 'ethnic minority' cultures. Rather, it means that the school must help such people to be subsumed within the school's ethos of competitive individualism.

By equating 'motivation' with willingness to compete for success on standardized tests, the Hargreaves Report diverts us from the profound issues of racism and pedagogy that underlie the officially recognized problem of 'underachievement'. In this way, the report individualizes the problem by attributing it to the 'underachieving' students; they are seen to suffer from cultural deficiencies rather than from a cultural devaluation by the school.

Furthermore, this monoculturalist approach has much in common with the 1960s integrationist perspectives, whose standing has since declined in favour of multicultural ones. Although multiculturalism can equally well evade the problem of racism, it is disturbing that a 1984 ILEA report on black underachievement simply ignores the entire debate on integrationist, multiculturalist and anti-racist perspectives. (See ILEA, *Race, Sex and Class* booklet no. 2, pp. 19–22 for commentary on perspectives emphasizing assimilation, diversity or equality, respectively.)

In conclusion, the Hargreaves Report diverts attention from the profound issues of racism and pedagogy that underlie the officially recognized problem of 'underachievement'. Thus the effects of racism get hidden as problems of the people who suffer from it.

SCIENCE 5–16: THE DES HOLDS THE LINE

In 1985 the DES issued a policy statement on how science education 'should have a place in the education of all pupils of compulsory school age, whether or not they are likely to go on to follow a career in science or technology.' For the foreseeable future these new guidelines will influence decision-making on the content and pedagogy of science education. Although the statement nowhere acknowledges any political purposes, we feel that its language involves implicit political choice. Our critique will expose and criticize these, in the hope of generating a real debate about the criteria and purposes of science education.

Before analysing the DES statement as such, it is worth noting that it was issued at a time of great public controversy about schooling. Many teachers and parents have been challenging racism in the schools. At the same time, right-wing pressure groups have been denouncing 'left-wing loony teachers' who supposedly indoctrinate students by dragging politics into the schools – as if schools were not always political anyway.

We recognize that the future of schooling will depend less upon a particular DES document than upon the course of that public controversy, innovative teaching methods, and so on. Nevertheless, a textual analysis can help clarify what images of science and what kinds of flexibility the DES is promoting.

IMAGES OF SCIENCE

What are, or should be, the aims of science education? The DES statement presents itself as a series of proposals for improving, and especially widening, science education, yet it offers little scope for debating what that means. The fundamental aims are not to be questioned, it seems, because science itself is not seen as problematic. The statement portrays science as benevolent, politically neutral and value-free: a force for good in society. It is to do with experiments in labs, forming hypotheses that can be tested in the classroom by applying some investigative skills. Accordingly, the statement promotes the narrow stereotype of the 'scientific method' as a purely technical search for the truth.

On that basis, it acknowledges a social connection in only one sense: 'Each of us needs to be able to bring a scientific approach to bear on the practical, social, economic and political issues of modern life' (para. 7). This language presumes a one-sided influence. It encourages the application of science to wider problems, while pretending that the science itself is purely technical, in no way shaped by social or political forces. Blindness to that connection is an integral part of scientists' training and of people's deference to scientific expertise. Thus this proposal reproduces the way such expertise is publicly invoked to foreclose any real debate about what our society could or should be.

Let us see two more examples in a similar vein:

> ... assessments of scientific knowledge applied in environmental, social and economic contexts should be based on the candidates' appreciation of the significance of scientific knowledge and understanding in the wider context: questions posed should require, to be answered successfully, appropriate scientific knowledge. (para. 65)

> The social and economic implications of scientific and technological activity also have a place, always provided that the teaching is essentially concerned to foster education in science itself. (para. 39)

Again, as regards applications or implications of science, priority is given to the technical aspects, with no attention given to the origins of scientific research – the 'science itself' – that turns out to lend itself to some applications rather than others. Teaching the 'science itself' is both an illusory neutrality and bad education.

A similar limitation arises in the DES's call for more integrated science courses. Such a move might open up greater possibilities for exploring wider aspects of science. However, when the DES suggests extending the integration beyond the conventional science courses, it does so in a one-sided way: that is, scientific approaches and principles are to be applied to other parts of the curriculum, since they 'also have aspects which can and should be treated scientifically' (para. 55). This seems to mean that aspects of the social world are to be understood in technical terms, but not the other way around.

'MAN'S PLACE IN THE WORLD'

Indeed, the DES statement proposes that science should encourage, among other things, 'an appreciation of a significant part of our cultural heritage and an insight into man's place in the world' (para. 9). That is, it does not suggest that our vision of a better society could lead us to reorient scientific research towards such aims – for example, by posing different questions for science to investigate. It suggests only that science can help us to understand the natural basis for the existing society – or rather, using traditional sexist language, the place of 'man' in the natural order. And what is 'our cultural heritage'? In the present context, this can only mean a nationalistic or European bias regarding what are the truly important contributions to science.

By comparison with those fundamental issues, the statement makes the minor concession that 'Just what knowledge of facts and principles should be taught is a matter for continual review, in the light of changes and developments in science and technology in the wider world' (para. 12). What about the changing social priorities that science embodies, or should do?

For example, surely an unprecedented nuclear disaster qualifies as grounds for a class to investigate why so many nuclear plants have gone wrong, why political decisions were taken to develop and build them, and why no more have been built in the USA since Three Mile Island.

These examples also demonstrate how dubious is the standard distinction between science and society, as well as between science and technology. As the DES rightly says, 'Links between teachers of science and CDT [Craft, Design and Technology] teachers are vital, if a damaging and unnecessary division between science and technology is to be avoided' (para. 57). However, if these links are based on the usual narrow definitions – scientific facts as related to technological things – the result is likely to confirm rather than challenge official models of our society as part of a natural order.

ABILITY LABELLING

Notwithstanding the DES's proposals for equal opportunity, the DES statement reproduces all the present divisions among students according to presumed ability: '. . . the most able . . . will proceed to further education and training in science and technology and may go on to make personal contributions to the nation's scientific effort' (para. 10). Clearly it treats science education as a national investment, to be made especially in those students who can offer a higher rate of return.

So, even though the DES statement heralds the advent of a more comprehensive science education for all pupils, it never questions why the system's priority is to select out a few prospective scientists, retrospectively labelled 'the most able'. Repeatedly using the language of supposed differences in innate ability, it accepts these as apparently natural divisions.

In particular, it remains satisfied with incorporating the 'less able' or 'weaker pupils' by giving them 'adequate opportunities to demonstrate, and to experience, success' (para. 66). By using 'differential approaches to assessment', this innovation would provide at best a second-rate 'success'. Of course, it is laudable

'OUR' CULTURAL HERITAGE

These statements are capable of so many interpretations they are likely to function simply as vague clichés (the fate of most aims, if they are not made specific and practical). Even without undertaking an analysis of what is meant by 'culture' there are many questions to be explored. Who are 'we' in reference to 'our' cultural heritage? Is it 'man' [sic], Europeans or Britons? Is the significance in relation to the two/three cultures of science, humanities and technologies, or in relation to ethnic groups, alternative civilizations? 'Man's place in the world' stated in such a universalistic way seems to imply a biological view, i.e. homo sapiens in relation to his/her environment and other species. What about different 'places in the world'? Is this what is hinted at by the contributions of other elements in the school curriculum? That could be interpreted as meaning that a study of man as a social being should be limited to the social sciences, that recognizably 'cultural' manifestations of belief, history, arts, etc. be studied only in those subject areas. Or is it really suggesting that the methods and competencies of science are cultural universals, so that we should be working out, say, the implications of hypothesizing, observing, etc., together with other teachers across the whole curriculum. This is a challenge which *would* take science teachers further into the realm of multicultural education. It is, however, a major task which is unlikely to be embarked on by most, if this is the clearest suggestion that the DES can give.

The impression given by the statement is: whatever happens elsewhere in schools and in the curriculum, science needs no cultural perspective, it is culture-free; and science teaching needs no anti-racist principle . . .

Martin Hollins on the DES policy statement, *Science News* 31 (Sept. 1985), p. 18.

to allow students 'to show what they can do rather than what they cannot do' (para. 13j), but this proposal would seem to formalize a hierarchy of presumed ability levels – for example, having the higher ones use abstract concepts and the lower ones take measurements. It would also serve to perpetuate the differences – cultural, class, gender – that children bring to their science classes.

EQUAL OPPORTUNITIES

The paragraphs on equal opportunities emphasize the 'failure of many girls to acquire a broad education in the main areas of science'. Is it really the girls' failure or the schools' failure? Of course the same problem applies to girls, black students and working-class students as a whole, albeit in slightly different ways. Why has the DES simply ignored the problem of working-class and black 'underachievement' in science, much less offered some explanation?

While prevalent teaching methods accentuate so-called 'ability' differences, co-operative learning methods could help to overcome the resulting exclusion and sense of failure, and could encourage a socially critical approach as well. The DES statement refuses to consider the possibility that individual assessment and the present organization of 'achievement' lie at the core of the problem. Perhaps that refusal is consistent with lamenting that 'the nation loses scientific and technological expertise', as the main preoccupation. In this context the DES can only mean a more equal opportunity for all pupils to become one of the few selected out as 'the most able' for 'success' in potentially making Britain once again great.

Lastly, while the DES proposes to offer some of every science subject to all pupils, it proposes to find the resources simply from expected savings due to 'falling rolls'. Thus the changes could come at the expense of those subjects (e.g. biology) which have seemed the least intimidating to students and offered them the greatest motivation from real interest in the subject. At the same time teachers would be expected to diversify into fields in which they have little training. Yet another irony for the much-vaunted equal opportunities.

CONCLUSION

The DES statement takes for granted the prevalent model of science teaching and proposes extending it to more students and more subjects – with some flexibility, but entirely within the present hidden agenda of science teaching. Lacking any proposals for teaching students about the forces that shape science and technology, or for overcoming presumed 'ability' divisions, it confirms the reduction of science education to a stratified skills training. It will be despite this DES statement that students, teachers and parents attempt to develop a real science education, worthy of the name.

INDEX

'ability' labelling 4, 10, 212, 214–17, 224
  DES statement 315–17
  as power relation 234
  as racism 233–64
aborigines 207
abortions 47, 276
abstract thought 236
'abuses' of science 16, 18, 43, 49
Africa 110, 111, 137
Agency for International Development (AID) 46–8
agenda-setting 25–8, 30
agribusiness 44, 49–51, 270, 277
agriculture 14, 29, 96, 137
Akamba people 138
aldicarb oxime (AO) 285–8
All London Teachers Against Racism and Fascism (ALTARF) 8
alternative world-views 31
American Civil War 193
American Medical Association 72, 73
Anthropological Society of London 190–2
anthropology 180, 190–3
anthropology of knowledge 79–80
applied science 302, 303
Arabs 303–4
Arnold, Dr Thomas 188
arts 34
assemblies 128
Assessment of Performance Unit (APU) 248–50, 256
assessment procedures 10, 210–64
  GASP 213–18, 219–22
  graded 219–31, 309
  see also 'ability' labelling
Association for Curriculum Development 11

Association for Curriculum Development in Science (ACDS) 10, 154, 295
automation 54–5, 251, 261

Bagehot, Walter 194
Banton, Michael 195
Baran, Grazyna vii
Barker, Martin 18
behavioural sciences 30
Bell Labs 25
Berger, Peter 78
Bernstein, Basil 251
Bhopal disaster 10, 56, 122, 270–92
  campaigns 277–9
  death toll 271, 284
biology 17, 90
  anti-racist 10, 124–35
  teaching 106–22
biotechnology 25, 51
birth control 96, 103
Black Panthers 62, 64, 68, 71, 73
Black Papers 246, 248
Black Parents Movement 253, 263
black pupils 107, 108, 111, 211
blood groups 118, 199, 200, 202–4
Blumenbach, Johann Friedrich 180, 182
Bonnet, Charles 179
breast cancer 20
breast milk 20
Britain 96–8
British Empire 178–9
Brittan, E. 129, 131
Bulldog 113, 114
Burt, Sir Cyril 18, 43, 120, 245, 247
Burton, Richard 190

Camper, Pieter 180–2
capitalism 3, 44, 115

Carlyle, Thomas 187, 192
cash crops 95, 100, 101, 104
Castleman, Barry vii, 272
Caucasians 180
Certificate of Secondary Education (CSE) 221, 222, 231, 262
cervical cancer 20
charities 28
Chemical Industries Association 292
chemistry 158–63
Chicano farm workers 50
circumcision 109
class analysis 125
class inequality 4
Coard, Bernard 235, 246
Cockcroft Report 172, 226
cognitive ability 241, 259
colonialism 2, 122, 176, 195
Combe, George 186
Commoner, Barry 45
communication skills 250
compensatory education 245
comprehensive education 147–51, 237–8, 246, 251
computer-aided design/manufacture (CAD/CAM) 23
computers 24, 56
containerization 23
contraception 20, 46–8
cosmetics 132
craniology 180–2
Crick, Francis 114
crop yields 104, 105
curriculum 4, 31–5
  anti-racist 5, 260–1
  change 148
  common-core 151–3, 159
  core 249
  development 160, 173
  innovation 156
  integrated 153
  multicultural 211
  status quo 4, 5
Curtin, P. D. 183
Cust, Robert Needham 196, 197
cybernetics 24

Darlington, C. D. 16
Darwin, Charles 84, 192, 193
Department of Education and Science (DES) 4, 11, 248, 312–18
diabetes 70
diets 20, 95, 97, 101
Dinham, Barbara vii, 271
disadvantage 210, 211, 245, 249, 258
division of labour 44, 76, 224, 240

Douglas, Mary 79–80
drinking water 109, 111
drug industry 28

East India Company 178
ecology 91
economism 81, 82
ecosystems 96, 137, 144
Edwards, W. F. 186
Ehrlich, Paul 45
elephant 139
11+ exam 237–8, 264
energy policies 14, 28, 44, 52, 159, 162
Engels, Friedrich 80, 81
English as a Second Language 128
equality 115
equality of opportunity 147, 148, 151–7, 252, 317
  definition 157, 174
ESN schools 246–7, 252, 255, 259
espionage 24
ethnic minority groups 297–9, 310
eugenics 195
evolutionism 188–90
examinations 4, 157, 213–31, 243
exchange rates 96
exploitation 2, 43, 117, 128, 130, 176
  British involvement in 122
Eyre, Governor 191–3
Eysenck, Hans 43, 113, 114, 119, 246

'facial angle' 180
famine 99–101, 122, 134, 138, 140
fascism 114, 115
fashions of science 18
fertilizers 101, 163
firewood 52
food 48, 50, 132, 163
  consumption 103–4
  preparation 112
food processing 50–1, 95, 97
Forman, Paul 79
Foucault, Michel 83
Friedson, Eliot 78
Fryer, Peter vii, 130, 176
funding of research 14, 18, 22, 24

Galton, Sir Francis 195
General Certificate of Secondary Examination (GCSE) 220–2, 226–31, 262
  Science Guide 226, 227, 229, 231

Index

genes 201–6, 241
    polymorphic 202–5
genetic counselling 19, 20, 69, 72
genetic engineering 20, 51
genetics 112–15, 199
genetic transplants 20
genocide 130
'geographical race' 199, 201
geography 128, 134, 308
Gill, Dawn vii
Gillie, Oliver 247
government policy 5, 295
Graded Assessments in Science Project (GASP) 212, 213–18, 219–22
grain production 103
grammar schools 246, 251
'Great Debate' 244, 248
Great World Depression 104
Green, Malcolm vii, 91
Green Revolution 29, 49, 97
Grunwick photo processing plant 54–5
Guttmacher, Alan F. 46

haemoglobin 59, 61, 64–5, 67, 68, 118
Harding, Jan 159, 160
Hargreaves, David 7, 295
Hargreaves Report 213–15, 256–7, 295, 307–11
health care 73–5
heart attacks 20
heredity 114, 198, 241, 244, 245, 259
Hessen, Boris 81
high-yield varieties (HYVs) of grain 49–50
history 128, 134, 308
HMI Report 131
Holland Park School 91, 147–74
    anti-racist education 154–6
    CDTMS faculty 152–4, 164
    common-core curriculum 151–3
    comprehensivization 148–51
    evaluation of pilot scheme 168–70
    'issues-based' science 156–68
    pedagogy 156
    reappraisal 166–74
    science department 153–4
home economics 128, 149, 152
hormones 25
Hunt, James 190–3

IBM 25
images of science 313–14

immigration 127, 245
immune system suppression 25
imperialism 2, 49, 122, 178–9
Improving Secondary Schools 7, 213, 295, 307
indentured servitude 44
India 50, 53
industrial action 159, 165, 166, 171, 174
'industrial-based planning' 23
infibulation 109
Inner London Education Authority (ILEA) 5, 128
    anti-racist policies 218, 225, 294
    anti-racist statement 296–300
    correspondence 301–6
    curriculum reform 5, 172
    graded assessment 219, 257
    teacher turnover 151
innovation in science 43–56, 148
insulin 70
integrated science courses 314
intelligence 119–20
    definition 119
    innate 223, 306
    racial differences in 43, 114
IQ 16, 30
    as capitalist science 235
    pedagogy for testing 237–8
    testing 18, 233–44
    debate 240–8, 258–9
    as power relation 239, 240
Ireland 99, 183
'issues-based' approach 156–63, 166–8, 171
ivory trade 138–49

Jamaica 191–3
Japanese 24, 200, 201
Jensen, Arthur 18, 43, 119, 244–6, 253
John, Gus 253
Johnston, Sir Harry 188
Jones, Tara viii, 271
Joseph, Sir Keith 5, 33, 34

Kames, Lord 182, 186
Kamin, Leon 245, 247
Keith, Sir Arthur 114
Kenya 136–46
Kidd, Benjamin 194
Kiernan, V. G. 184
Knox, Dr Robert 189, 191
Krebs, Sir Hans 17
Kuhn, Thomas 79
kwashiorkor 108, 134

labour 14, 43
labour process studies 83
Lamb, Charles 196–7
land 101
land rights 144, 145
Lawrence, Sir William 184
Levidow, Les viii, 14, 212
Lewontin, Richard viii, 177
life expectancy 26
Lindsay, Liz viii, 90
Linnaeus 179
literacy 248
Long, Edward 178–9
Lorenz, Konrad 17, 113–14, 121
Lusigi, W. J. 136–8, 144, 145

MacFarlane Report 251
malaria 19, 25, 51, 65, 66
malnutrition 26, 51, 71, 95–105
Mannheim, Karl 78
Manpower Services Commission (MSC) 248, 250, 251, 261, 262
marketing 20, 22
Marx, Karl 80
Masai 140, 143, 200, 201
mass media 127
mathematical ability 120
Medawar, Sir Peter 25
mediation theory 80–2
medical care 63, 64, 73–5
medicine 28–9
Merck, Sharp and Dohme 292
Merton, Robert K. 78
methylene chloride 285, 288
methyl isocyanate (MIC) 271–6, 278–81, 285–90
Michaelson, Michael viii, 15
microelectronics 14, 44, 53–6
migration 119
Mill, John Stuart 187, 192
mixed-ability classes 148, 152, 230, 251–2
Monday Club 5
Montagu Ashley 121
mortality rates 26–7
Morton, Samuel George 190
motivation 215, 219, 256, 309, 310
multicultural education 5, 91, 124–35
multiculturalism 5, 107, 124–35, 211, 259–61, 311
multinational companies 23, 44
Murray, Gilbert 188
museums 30, 34

National Front 113, 114
nationalism 125
nationality laws 1
National Parks 139, 141–6
natural resources 14, 43
neutrality of science education 4, 90
Nixon, Richard 63, 64, 71, 75, 245
non-stick frying pans 23
Northern Ireland 55
nuclear physics 24
nuclear power 52, 159
Nuffield Advanced Biology Course 113, 115, 118
Nuffield Physics Project 228
numeracy 248
nutrition 10, 95–105

oceanography 22
oppression 109, 127, 147, 260
On the Origin of Species 84, 194

particle physics 305
Patel, Vinod viii
patronage 28–30
Pauling, Linus 65
PCBs 291–2
peace studies 7, 33, 34
Pearson, Karl 195
pensions 102
People's Free Medical Centres 73–5
pesticides 101, 270, 277, 279
    carbaryl 282
    MIC-based 289
    Temik 285
philosophy of science 30–1
phrenology 184–6, 190, 194
physical traits 177, 198
Pim, Commander Bedford 192, 193
police 55, 56, 124
political priorities 14
pollution 109–10, 282, 292
polytechnics 34
population control 14, 44–9
positivism 32
poverty 44, 46, 49, 90, 134
powdered milk 20, 22
practical work 166, 214
Prichard, James Cowles 182
'primary race' 119
priorities of science 18, 20–5
process skills 215, 219–25, 229
profiling 250–2, 309
proteins 96, 97, 101, 201–2, 204
    monomorphic 202
    polymorphic 202
public health 19

Index

Puerto Rico 25, 30, 46–8
pupil/teacher ratio 153
pure science 301–3, 306

quantified assessment 217
questions 19

Race Relations Act (1976) 298
racial differentiation 198–207
racial discrimination 17, 298
racial extinction 195
'racial purity' 118–19
racial stereotypes 3, 234, 239, 300
racism
   'ability' labelling as 233–64
   in black societies 124
   consensual 196–7
   definition 124
   Hargreaves Report 309–11
   mythology 176
   overt 112
   pseudo-scientific 176–97
   roots of 108, 124, 125
   in scientific innovation 42–56
   structural 3
Radical Science Collective 8
radioactivity 53
Rampton Report 125, 126, 254–6
rare-earth metals 23
red blood cells 19, 59, 64
Redbridge report 252, 253
rickets 98, 108, 114
'right to know laws' 279, 282
Rockefeller charities 28–9, 49
Roebuck, John Arthur 195

Scance, General Larry 23
schistosoma 109, 111
Schools Council 131, 134
Science 5–16 11, 295, 312–18
Scott, Dr Roland B. 63, 75
screening 19, 20, 69, 71
Secondary Mathematics Individualized Learning Experiment (SMILE) 171
'secondary race' 119
seismology 22
Sethi, Ashok viii
sexism 4, 75, 132, 314
sickle cell anaemia 15, 19, 59–75
   case histories 66–8
   characteristics 59–60, 64
   as 'interesting pathology' 61
   research 19, 20, 51, 61–4, 71
   sudden deaths 66–9
'sickle cell trait' 62, 65–6, 68

silicon chips 53
Singh, Birendra ix, 212
Singh, Europe ix
skin colour 133, 207
skulls 180, 184, 186, 190
slavery 44, 138, 178, 188, 192, 193
Smith, Henry ix
Social Darwinism 193–6
Society for Participatory Research (PRIA) 277–9
socioeconomic status 241–2
sociology of education 120
sociology of knowledge 78–9, 84–5
sociology of science 76–8
Soemmerring, Thomas 182
South Africa 159
'special needs' 261–3
Spencer, Herbert 192, 194
Spice, Dr John 302–4, 306
sterilization 46–8, 245
streaming 230, 246, 251, 300
structuralist studies 83–4
superiority 201–6
Swann Report 211, 257–9
systems theory 24

Tasmania 195–6
teacher turnover 151
Technical and Vocational Education Initiative (TVEI) 251, 262
teleology 187–8
television 136, 176
textbooks 115–18, 166, 172, 173
Third World
   exploitation 116
   impoverishment of 2, 44
   negative image of 108
   resources 23
Three Mile Island nuclear disaster 271, 315
tomato industry 50
Tomlinson, Sally 247, 255
tourism 91, 142–3
Townsend, H. 129, 131
trade unionism 17, 50, 55, 274, 276, 277
traditional culture 144–6
transnational corporations 124

underachievement 211, 213, 214, 219, 242, 263
   causes 249, 252–4, 256
   definition 210
   Hargreaves Report 307–11
   by West Indians 125–6, 254–5, 258

underemployment 55, 56, 224, 251
unemployment 55, 56, 104, 115
  among school leavers 219, 224, 251
UNESCO 115, 136
Union Carbide Corporation 10, 270–92
  double standards 280–2
  safety record 274, 280–2, 285–7
  US plant 271, 279, 284, 285, 289, 290
United States 19, 22, 23, 50, 279
United States Air Force 23

Vance, Michael ix, 90, 94
variation in humans 118–19, 198–207
vegetarian diet 97
Verity, Robert 186

vitamins 25
vocational training 261–3

wage-labour 44
Waterman, Professor Talbot 22
Wechsler test 238
West Indians 125–6, 254–5, 258–9
White, Charles 182
wildlife conservation 10, 136–46
Willey, R. 130
'worksheet' approach 164, 166
World War 2 198

Young, Robert M. ix, 14, 15
Youth Opportunities Programme 262
Youth Training Scheme (YTS) 55, 262